Lecture Notes in Mathematics

Edited by A. Dold and B. Eckmann

1202

Arne Dür

Möbius Functions, Incidence Algebras and Power Series Representations

Springer-Verlag

Berlin Heidelberg New York London Paris Tokyo

Author

Arne Dür
Institut für Mathematik der Universität Innsbruck
Technikerstr. 15, A-6020 Innsbruck, Austria

Mathematics Subject Classification (1980): 05A10, 05A15, 06A10, 13J05, 14L15, 16A45

ISBN 3-540-16771-4 Springer-Verlag Berlin Heidelberg New York
ISBN 0-387-16771-4 Springer-Verlag New York Berlin Heidelberg

Printed in Germany

Printing and binding: Beltz Offsetdruck, Hemsbach/Bergstr.
2146/3140-543210

INTRODUCTION

Möbius inversion, both on locally finite partially ordered sets (popularized by G.C.Rota [Ro,64]) and on monoids with the finite factorization property (introduced by P.Cartier and D.Foata [CF,69]), has become an important area of combinatorics. In case of partially ordered sets, the historical development and recent progress is traced in an article of C.Greene [Gr,82], and applications to enumeration theory have been surveyed by E.A.Bender and J.R.Goldman [BG,75]. The algebra wherein Möbius inversion takes place, is usually called the "incidence algebra".

In this book, a general setting for incidence algebras is proposed which contains almost all examples from the literature and also admits a substantial structure theory. It is written for people interested in combinatorics and algebra, and is influenced by papers of G.C.Rota, R.P.Stanley, P.Leroux, A.Joyal and others ([DRS,72],[JR,79],[CLL,80], [Joy,81]). In a joint work of U.Oberst and the author [DO,82], the present theory of incidence algebras has been sketched.

Our construction of the incidence algebra from a combinatorial category is inspired by the idea of P.Hall who used certain numbers of subgroups of abelian p-groups as structure coefficients of an algebra ([Ha,59],[Ma,79]). As a structural result, we show that the invertible elements of the incidence algebra define a triangulable affine group ([DG,70]) over the integers. For special combinatorial categories, the incidence algebra becomes even a cocommutative bialgebra, and is the covariant bialgebra of the affine monoid of "multiplicative functions". Particular attention is paid to the question when the incidence algebra (or the affine monoid of multiplicative functions) can be realized as an algebra of formal power series (or as an affine monoid of transformations in a power series algebra, resp.). Such a representation connects the count in the category with the multiplication (or the substitution, resp.) of special power series, and gives rise to power series identities. Here we pursue the idea of P.Doubilet, G.C.Rota and R.Stanley in their article "The idea of generating function" [DRS,72], an exposition of which can be found in the book of M.Aigner [Ai,79]. This idea occurs also in the book of I.G.Macdonald [Ma,79] where the Hall algebra is mapped isomorphically onto the algebra of symmetric functions in infinitely many variables, thus relating the subgroup lattices of abelian p-groups to the multiplication rule of the Hall-Littlewood functions. Finally, we work out various examples of combinatorial interest, thereby meeting different kinds of triangulable (and unipotent) affine groups. In most examples, we establish a faithful representation by power series which can be used to compute Möbius functions or to set up generating functions for enumeration problems.

After this general survey I give a more detailed description.

In chapter I we introduce the concept of a "categorical structure", construct the incidence algebra and develop its theory. A categorical structure is a triple $\underline{K}, M, \sim$ where $\underline{K}$ is a category, M is a class of morphisms in $\underline{K}$ and $\sim$ is an equivalence relation on M with certain properties. The equivalence classes t in $T=M/\sim$ are called types. Given a commutative ring k, let k^T be the topological k-module of functions $f:T\to k$, endowed with the componentwise linear operations and the product topology. The convolution product of $f,g \in k^T$ is defined by

$$fg(t) = \Sigma_{t_1,t_2 \in T}\, G(t;t_1,t_2)f(t_1)g(t_2) \quad , \quad t\in T \quad ,$$

where the "section coefficient" $G(t;t_1,t_2)$ counts how often a morphism of type t can be factorized into a morphism of type t_1 and a morphism of type t_2, up to simplicial equivalence. With this multiplication, k^T becomes an associative topological k-algebra with unit, called the incidence algebra $k(T;G)$. The Möbius function μ in $k(T;G)$ is the inverse of the zeta function $\zeta:T\to k$, $\zeta(t)=1$ for all t. As special cases, the incidence algebras of locally finite posets and of monoids with the finite factorization property are obtained. The incidence algebra $k(T;G)$ possesses considerable structure. The dimension of morphisms in M, defined as the maximal length of factorizations in M, induces a filtration

$$k(T;G)=I(0) \supset I(1) \supset I(2) \supset \dots$$

by closed two-sided ideals converging to $\{0\}$ such that

$$I(d_1)I(d_2) \subset I(d_1+d_2) \quad \text{for all } d_1,d_2 \ .$$

This implies that the invertible elements of $R(T;G)$, R a commutative k-algebra, form a triangulable affine group Ω. In the literature, similar constructions of incidence algebras have been considered. From [CLL,80] we differ by counting 2-simplices up to simplicial equivalence, but compare [Joy,81]. In [He,72],[JR,79],[BBR,80] and [Joy,81] one starts with numbers $G(t;t_1,t_2)$ ($t,t_1,t_2 \in T$, an arbitrary set) which satisfy special identities ensuring that the convolution product is well-defined, associative and has a unit. This approach is more general, but little can be said on the structure of the resulting algebras. Furthermore, in most of our examples, there are no explicit expressions for the section coefficients whereas the conditions for a categorical structure are easily verified.

Our approach is intended primarily for the study of the partially ordered sets of subobjects (or quotient objects) in a combinatorial category, such as the categories of finite sets, finite vector spaces, words, finite abelian groups, finite sets under group action, rooted forests, matroids, partitions etc. Suppose now that M is a class of monomorphisms (Dual results are obtained for a class of epimorphisms). Given an object Y in $\underline{K}$, let $Sub_M(Y)$ denote the poset of subobjects of Y in M, and let $I(Sub_M(Y),k)$ be its incidence algebra over the ground ring k. Then, as a central result, there exists a continuous k-algebra homomorphism

$$l_Y: k(T;G) \to I(Sub_M(Y),k)$$

which preserves the zeta and Möbius function. For instance, let $\underline{K}$ be the category of finite abelian groups, and define an equivalence relation on the class M of all monomorphisms by $(s_1:X_1 \to Y_1) \sim (s_2:X_2 \to Y_2)$ if and only if the factor groups $Y_1/s_1(X_1)$, $Y_2/s_2(X_2)$ are isomorphic. Here the posets $Sub_M(Y)$ are the subgroup lattices of finite abelian groups, and the incidence algebra $\mathbb{C}(T;G)$ appears already in an article of S.Delsarte [Del,48] where the Möbius function $\mu \in \mathbb{C}(T;G)$ is computed. Specializing to abelian p-groups, we get another incidence algebra which contains the Hall algebra ([Ma,79]) as a subalgebra.

The problem when the incidence algebra of a categorical structure is isomorphic to the large monoid algebra of a monoid N with the finite factorization property (in particular, a power series algebra) or to the incidence algebra of a poset P, is shown to be related to the second cohomology monoid of N or P resp. We point out that even $H_n^2(\mathbb{N}_o,\mathbb{Q})$ is difficult to characterize, and apply some results on second cohomology groups to incidence algebras.

We finish chapter I by a series of examples of incidence algebras taken from the literature.

Chapter II introduces Krull-Schmidt categories. Essentially, these are categories where the theorem of Krull-Schmidt ([At,56]) holds: Every object in $\underline{K}$ admits a direct sum decomposition into finitely many indecomposable objects which is unique up to isomorphism.
Many combinatorial categories have this property, e.g. the categories of finite sets, finite vector spaces, graphs, trees, finite sets under group action, matroids etc. We prove two exponential formulas which

are power series identities of the form $a = \exp(b)$. The first one generalizes the classical generating function for the number of partitions of finite sets to arbitrary Krull-Schmidt categories with finite automorphism groups. The second one relates, loosely speaking, the count of all objects over a given base to the count of indecomposable objects, and, for instance, can be used to set up a generating function for the number of roots of elements in the symmetric groups. In the combinatorial literature, "exponential formulas" have been studied in [FS,70],[F,74],[BG,71],[St,78],[St,78'] and [Joy,81].

In chapter III we consider special categorical structures $\underline{K}$,M,~ whose significant features are the following: $\underline{K}$ is a Krull-Schmidt category where the decomposition of any object into finitely many indecomposable objects is unique up to the order, and M satisfies the "sheaf condition": Given a morphism $s:X\to Y$ in M, there exists, for any partition π of Y, a unique partition σ of s which induces on Y the partition π. E.g., in the category of finite sets, for $\pi=\{B_i;i=1,..,l\}$, we have $\sigma=\{s_i:s^{-1}(B_i)\to B_i;i=1,..,l\}$ where the s_i are the restrictions of s. More generally, all morphisms in toposes (i.e. categories of sheaves, [GV,72]) have this property. Therefore, we call such categorical structures <u>sheaflike</u>.

With the addition induced from the direct sum on M, the set of types T is a commutative monoid and gives rise to a topological coalgebra structure on k^T which is compatible with the algebra structure of k(T;G) in virtue of the sheaf condition. The cocommutative topological bialgebra k(T;G) is called the <u>incidence bialgebra</u>, and is the covariant bialgebra of the affine monoid Mu of <u>multiplicative functions</u> on T. Here, for any commutative k-algebra R, Mu(R) is the set of functions $f:T\to R$ such that $f(0)=1$ and $f(v+w)= f(v)f(w)$ for all $v,w\in T$. In particular, the zeta and the Möbius function are multiplicative. As a closed subgroup of the triangulable group Ω, the affine group E(Mu) of invertible multiplicative functions is also triangulable and contains the <u>unipotent</u> normal subgroup Mu" as the algebraically interesting part. The Lie algebra of Mu" is <u>topologically nilpotent</u> and determines the isomorphism type of Mu" over the rationals. Remarkably, the structure coefficients of this Lie algebra with respect to a distinguished topological basis are section coefficients of a very special form. In the literature ([JR,79],[Joy,81]), bialgebras have been constructed from numbers $G(t;t_1,t_2)$ ($t,t_1,t_2\in T$, a commutative monoid) which in addition to the identities mentioned above satisfy

$$G(v+w;t_1,t_2) = \Sigma_{v_1+w_1=t_1,v_2+w_2=t_2} G(v;v_1,v_2)G(w;w_1,w_2)$$

for all $v,w,t_1,t_2 \in T$. In our approach, this Vandermonde-like identity follows from the sheaf condition.

Combinatorialists will be interested into representations of the affine monoid Mu by endomorphisms of power series algebras in order to set up generating functions. Unfortunately, we have no global existence theorem on faithful power series representations of the monoid Mu, but some results in a special situation and various examples. The shape of power series representations of Mu in a finite number of variables can be described using the fact that E(Mu) is triangulable. In differential geometry there is a realization theorem for transitive Lie algebras by derivations of power series algebras ([Go,72]), and we hope that someone will prove an analogous result in our case.

At the end of chapter III we discuss the two basic examples. For the subset lattices of finite sets, the unipotent group Mu" is isomorphic to the one-dimensional additive group G_a, $G_a(R) = R$ with +, while for the partition lattices of finite sets Mu" is isomorphic to the infinite-dimensional group of automorphisms of the power series algebra in one indeterminate z mapping z to z + terms of higher order.

In chapter IV we work out more intricate examples which, as I believe, prove the usefulness of the present techniques for combinatorics. In §1,2,5,6 we establish faithful power series representations of the affine monoids of multiplicative functions which are used to calculate the Möbius functions of special posets or to solve enumeration problems connected with them. For instance, in §2 we count invariant partitions of finite sets under group action, in §5 we compute the Möbius function of the set of partitions of {1,..,n} having only blocks of odd size, and in §6 we determine, for a finite field $\mathbb{F}$, the distribution of the number of eigenvalues in $\mathbb{F}$ of an $m \times m$-matrix whose entries are chosen from $\mathbb{F}$ at random. In §3,4,7 we review and extend three examples from the literature. §3 gives an application of the unipotent group derived from rooted forests to the numerical integration of ordinary differential equations which is due to J.C.Butcher ([But,72],[HW,74]). In §4 it is shown, following [JR,79], how special classes of matroids give rise to graded Hopf algebras. Finally, in §7, we rapidly calculate from a faithful power series representation the cardinalities, Whitney numbers, Möbius functions and characteristic polynomials of a class of geometric lattices studied by T.A.Dowling ([Dow,73'],[Han,84]).

References to the bibliography are by the first letters of the name(s) of the author(s) and by the last two digits of the year of publication. In addition to the papers quoted in the notes, I have included a list of recent articles connected with combinatorial Möbius inversion.

This book was written at the Institut für Mathematik of the University of Innsbruck. I wish to thank Professor U. Oberst for introducing me into the subject and for encouraging me during my studies. Thanks are also due to the organizers of the Seminaire Lotharingien de Combinatoire for the opportunity to lecture on these materials in several meetings. I am grateful to the Austrian Ministry of Science for supporting a literature search through a research grant.

March, 1985 Arne Dür

NOTATION AND CONVENTIONS

$\mathbb{N}_o$	the set of natural numbers including 0
$\mathbb{N}$	the set of natural numbers (excluding 0)
$\mathbb{Z}$	the ring of integers
$\mathbb{Q}$, $\mathbb{R}$, $\mathbb{C}$	the field of rational, real or complex numbers
$\emptyset$	the empty set
$\#X$	the cardinality of a finite set X
$X-Y$	the complement of the subset Y in the set X
$\dot{\cup}_{i\in I} X_i$	the disjoint union of the sets X_i
$\oplus_{i\in I} V_i$	the direct sum of the commutative monoids or modules V_i
$E(G)$, $E(R)$	the group of invertible elements of a monoid G, ring R

$\mathbb{N}_o(I) = \{n\in(\mathbb{N}_o)^I;\ n(i)\neq 0 \text{ for only finitely many } i\in I\}$,

endowed with the componentwise addition, is the free commutative monoid generated by the index set I.

$\varepsilon(i) = (0\ldots010\ldots0)$, 1 at the position i .

$(\varepsilon(i);i\in I)$ is the standard basis of $\mathbb{N}_o(I)$.

For $n\in \mathbb{N}_o(I)$,

$|n| = \Sigma_{i\in I}\, n(i)$ (in fact, a finite sum) and

$n! = \Pi_{i\in I}\, n(i)!$ (in fact, a finite product) .

If G is a commutative monoid with unit 1 and if $(g_i)_{i\in I} \in G^I$, then

$g^n = \Pi_{i\in I}\, (g_i)^{n(i)}$ (By convention, $h^0=1$ for all $h\in G$) .

G often is the multiplicative monoid of a commutative ring.

CONTENTS

CHAPTER I

CATEGORICAL STRUCTURES AND INCIDENCE ALGEBRAS

§1. Categorical structures

Let $\underline{K}$ be a category. We assume that $\underline{K}$ is skeletal-small, i.e. that the isomorphism classes of objects in $\underline{K}$ form a set.
Let M be a class of morphisms in $\underline{K}$ with the following property:

(M1) M contains all isomorphisms in $\underline{K}$ and is closed under composition.

(1.1) DEFINITION: For $l=1,2,3,..$, let
$$S_l(M) = \{(s_1,..,s_l)\in M^l;\ \text{the composition } s_l..s_1 \text{ exists}\}$$
be the class of <u>l-simplices</u> of M (in particular, $S_1(M)=M$). We define an equivalence relation on $S_l(M)$, called <u>simplicial equivalence</u>, by $(s_1,..,s_l)\approx(r_1,..,r_l)$ if and only if there are isomorphisms $a_1,..,a_{l-1}$ such that the diagram

$$\begin{array}{ccccccccccccc}
X_0 & \xrightarrow{s_1} & X_1 & \xrightarrow{s_2} & X_2 & \xrightarrow{s_3} \cdots \xrightarrow{s_{l-2}} & X_{l-2} & \xrightarrow{s_{l-1}} & X_{l-1} & \xrightarrow{s_l} & X_l \\
\downarrow{\scriptstyle id} & & \downarrow{\scriptstyle a_1} & & \downarrow{\scriptstyle a_2} & & \downarrow{\scriptstyle a_{l-2}} & & \downarrow{\scriptstyle a_{l-1}} & & \downarrow{\scriptstyle id} \\
Y_0 & \xrightarrow{r_1} & Y_1 & \xrightarrow{r_2} & Y_2 & \xrightarrow{r_3} \cdots \xrightarrow{r_{l-2}} & Y_{l-2} & \xrightarrow{r_{l-1}} & Y_{l-1} & \xrightarrow{r_l} & Y_l
\end{array}$$

commutes, i.e. $a_1s_1=r_1$, $a_is_i=r_ia_{i-1}$ for $i=2,..,l-1$ and $s_l=r_la_{l-1}$ (For $l=1$ simplicial equivalence is just the identity relation). We denote the equivalence class of the l-simplex $(s_1,..,s_l)$ by $\langle s_1,..,s_l\rangle$, and the class of all equivalence classes of l-simplices of M by
$$S_l\langle M\rangle = S_l(M)/\approx \quad .$$

We assume:

(M2) For every 2-simplex $(s_1,s_2)\in S_2(M)$, the only isomorphism a such that the diagram

$$\begin{array}{ccccc}
 & & X_1 & & \\
 & \overset{s_1}{\nearrow} & & \overset{s_2}{\searrow} & \\
X_0 & & \downarrow{\scriptstyle a} & & X_2 \\
 & \underset{s_1}{\searrow} & & \underset{s_2}{\nearrow} & \\
 & & X_1 & &
\end{array}$$

commutes, is the identity on X_1.

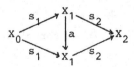

(1.2) DEFINITION: Let s be a morphism in M. A <u>factorization</u> of s of length l is an l-simplex $(s_1,..,s_l)\in S_l(M)$ where no s_i is an isomorphism, such that $s=s_l..s_1$. It is called <u>proper</u> if $l\geq 2$.

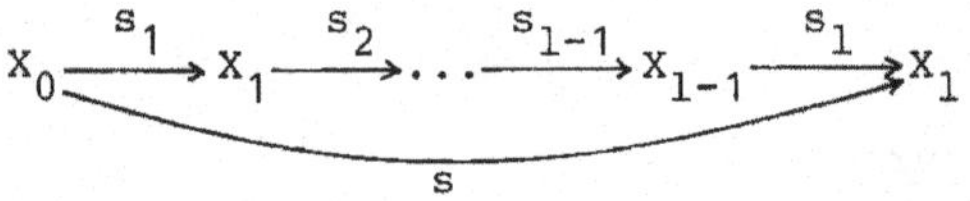

Observe that any l-simplex equivalent to $(s_1,..,s_l)$ also is a factorization of s. The factorization $(s_1,..,s_l)$ is <u>maximal</u> if no s_i can be factorized properly. The <u>dimension</u> of s is the least upper bound in $\mathbb{N}_0\cup\{\infty\}$ of $\{l;(s_1,..,s_l)$ a factorization of $s\}$, and will be denoted by dim(s).

(1.3) PROPOSITION: Equivalent are

(i) Every morphism in M has, up to simplicial equivalence, only finitely many factorizations.

(ii) All morphisms in M have finite dimension and, for every $s\in M$, the set $\{\langle s_1,s_2\rangle\in S_2\langle M\rangle; s_2s_1=s\}$ is finite.

Proof: Clearly (i) implies (ii) because, for $\langle s_1,s_2\rangle\in S_2\langle M\rangle$ with $s_2s_1=s$, $\langle s_1,s_2\rangle=\langle id,s\rangle$ if s_1 is an isomorphism, and $\langle s_1,s_2\rangle=\langle s_1,id\rangle$ if s_2 is an isomorphism. To prove the other direction, we show that, for $l\geq 3$ and $s\in M$,

$$\text{the set } \{\langle s_1,..,s_l\rangle\in S_l\langle M\rangle; s_l..s_1=s\} \text{ is finite .}$$

We proceed by induction on $l\geq 2$, and start with (ii). Let $R\subset S_2(M)$ be a system of representatives of $\{\langle p,q\rangle\in S_2\langle M\rangle; qp=s\}$. Because of (ii) R is finite. For any 2-simplex $(p',q')\in S_2(M)$ with $q'p'=s$ there is a unique $(p,q)\in R$ such that $\langle p',q'\rangle=\langle p,q\rangle$, and, by (M2), a unique isomorphism a such that $ap'=p$ and $q'=qa$.

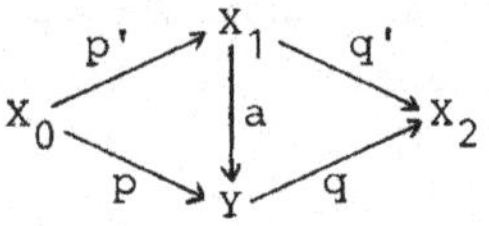

Thus we have a map

$$\begin{array}{ccc} \{\langle s_1,..,s_l\rangle\in S_l\langle M\rangle; s_l..s_1=s\} & & \langle s_1,..,s_l\rangle \\ \downarrow F & & \downarrow \\ \{((p,q),\langle r_1,..,r_{l-1}\rangle)\in R\times S_{l-1}\langle M\rangle; r_{l-1}..r_1=p\} & & ((p,q),\langle s_1,..,s_{l-2},as_{l-1}\rangle) \end{array}$$

where p,q and a are defined as above for $p'=s_{l-1}..s_1$ and $q'=s_l$. Obviously, F is a bijection whose inverse is given by $F^{-1}((p,q),\langle r_1,..,r_{l-1}\rangle)=\langle r_1,..,r_{l-1},q\rangle$.
The codomain of F is the disjoint union of the sets
$\{(p,q)\}\times\{\langle r_1,..,r_{l-1}\rangle\in S_{l-1}\langle M\rangle; r_{l-1}..r_1=p\}$, $(p,q)\in R$,
which are finite by the induction hypothesis. Therefore, the codomain and the domain of F are finite. □

The last assumption on M is

(M3) Every morphism in M has, up to simplicial equivalence, only finitely many factorizations.

(1.4) LEMMA: For $s \in M$, $\dim(s)=0$ if and only if s is an isomorphism.

Proof: If s is no isomorphism, then (s) is a factorization of s and $\dim(s) \geq 1$. Conversely, suppose that s is an isomorphism and that $\dim(s) \geq 1$. Let $(s_1,..,s_l)$ be a factorization of s of length $l \geq 1$. From $id=s_l..s_1 s^{-1}$ it follows that $id=s_l..s_2(s_1 s^{-1})s_l..s_2(s_1 s^{-1})s_l...$ yielding factorizations of id of arbitrary length. But this contradicts $\dim(id)<\infty$. □

Finally, let ~ be an equivalence relation on M with the properties (~1)-(~4). We call ~ <u>equivalence</u> and its equivalence classes <u>types</u>. The type of a morphism $s \in M$ will be denoted by $\bar{s}$.

(~1) Isomorphy in the category of morphisms in <u>K</u> implies equivalence: If s,r are morphisms in M such that bs=ra where a,b are isomorphisms, then s~r.

$$\begin{array}{ccc} X & \xrightarrow{s} & Y \\ a\downarrow & & \downarrow b \\ Z & \xrightarrow{r} & W \end{array}$$

(~2) The isomorphisms in <u>K</u> are saturated with respect to ~: If a is an isomorphism and $s \in M$ such that a~s, then s too is an isomorphism.

(~3) If $s:X \to Y$ and $r:Z \to W$ are morphisms in M, then s~r implies that $id_X \sim id_Z$ and $id_Y \sim id_W$.

The most important condition for ~ is

(~4) For any types t,t_1 and t_2, the number
$G(t;t_1,t_2) = \#\{\langle s_1,s_2\rangle \in S_2\langle M\rangle ; s_2 s_1=s, \overline{s_1}=t_1, \overline{s_2}=t_2\}$
is independent of the choice of the representative $s \in t$.

From (~1)-(~4) it follows that the types form a set

$$T = M/\sim$$

which can be decomposed into the disjoint subsets

$T' = \{\bar{s}; s \text{ an isomorphism}\}$ and $T(1) = \{\bar{s}; s \in M, \text{but no isomorphism}\}$.

Moreover, we have maps

dom: $T \to T'$, $\overline{s:X \to Y} \mapsto \overline{id_X}$, and cod: $T \to T'$, $\overline{s:X \to Y} \mapsto \overline{id_Y}$.

The numbers $G(t;t_1,t_2)$ are called <u>section coefficients</u>. In the sequel, we will show that the $G(t;t_1,t_2)$ are "section coefficients" in the sense of G.C.Rota ([JR,79],pp.96; compare also [He,75] and [BBR,80]).

(1.5) The <u>standard example</u> of equivalence is isomorphy in the category of morphisms in $\underline{K}$:

Set $s\sim r$ if and only if there are isomorphisms a,b such that $bs=ra$.

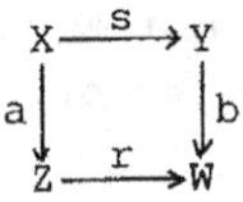

The properties (~ 1)-(~ 3) are obvious, and for (~ 4) observe that

$$\{\langle s_1,s_2\rangle \in S_2\langle M\rangle ; s_2s_1=s, \overline{s_1}=t_1, \overline{s_2}=t_2\} \qquad \langle s_1,s_2\rangle$$
$$\downarrow \qquad\qquad \downarrow$$
$$\{\langle r_1,r_2\rangle \in S_2\langle M\rangle ; r_2r_1=r, \overline{r_1}=t_1, \overline{r_2}=t_2\} \qquad \langle s_1a^{-1}, bs_2\rangle$$

is a bijection.

(1.6) DEFINITION: We call a triple $\underline{K},M,\sim$ where $\underline{K}$ is a skeletal-small category, M is a class of morphisms in $\underline{K}$ with (M1)-(M3) and $\sim$ is an equivalence relation on M with (~ 1)-(~ 4), a <u>categorical structure</u>.

The concept of categorical structure will be justified by the various examples it includes. In the rest of this section, we work out some properties of the section coefficients $G(t;t_1,t_2)$.

(1.7) LEMMA: Let t be a type. Then $G(t;t_1,t_2)\neq 0$ only for finitely many $t_1,t_2 \in T$.

Proof: Choose an $s\in t$. Because of (~ 1) we can define a map $\{\langle s_1,s_2\rangle \in S_2\langle M\rangle ; s_2s_1=s\} \to \{(t_1,t_2)\in T^2 ; G(t;t_1,t_2)\neq 0\}$, $\langle s_1,s_2\rangle \mapsto (\overline{s_1},\overline{s_2})$. Since it is surjective, the assertion follows from (M3). □

(1.8) LEMMA: Let t,t_1,t_2 be types with $G(t;t_1,t_2)\neq 0$.

(i) If $t_1\in T'$, then $t_1=\mathrm{dom}(t)$, $t_2=t$ and $G(t;t_1,t_2)=1$.

(ii) If $t_2\in T'$, then $t_1=t$, $t_2=\mathrm{cod}(t)$ and $G(t;t_1,t_2)=1$.

Proof: (i) Fix a representative s of t. For any $\langle s_1,s_2\rangle \in S_2\langle M\rangle$ with $s_2s_1=s$, $\overline{s_1}=t_1\in T'$ and $\overline{s_2}=t_2$, s_1 is an isomorphism and $\langle s_1,s_2\rangle = \langle id,s\rangle$ since the diagram

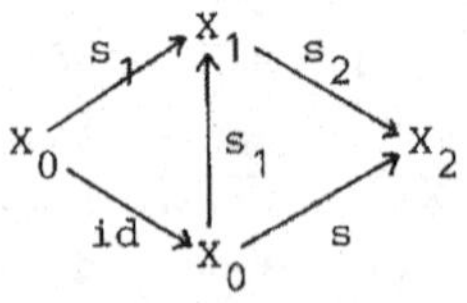

commutes. Hence $t_1=\overline{s_1}=\overline{id}=\mathrm{dom}(t)$, $t_2=\overline{s_2}=\overline{s}=t$ and $G(t;t_1,t_2)=\#\{\langle s_1,s_2\rangle \in S_2\langle M\rangle ; s_2s_1=s, \overline{s_1}=t_1, \overline{s_2}=t_2\}=1$.

(ii) is proved by duality: Instead of $\underline{K}$ consider the dual category $\underline{K}^{op}$ where the arrows are reversed. Obviously, $\underline{K}^{op}$,M and ~ satisfy all conditions and $G^{op}(t;t_1,t_2)=G(t;t_2,t_1)$. Thus, applying (i) to $\underline{K}^{op}$,M,~ we obtain (ii). □

(1.9) THEOREM: Let l be a positive integer and let $t,t_1,..,t_l$ be types. Then the number

$$G(t;t_1,..,t_l) = \#\{<s_1,..,s_l>\in S_l<M>;s_l..s_1=s,\bar{s}_1=t_1,..,\bar{s}_l=t_l\}$$

is independent of the choice of the representative s∈t. Of course,

$$G(t;t_1)=1 \quad \text{if } t=t_1 \quad \text{and} \quad G(t;t_1)=0 \quad \text{otherwise} .$$

When l≥3, we call the numbers $G(t;t_1,..,t_l)$ <u>higher section coefficients</u>. They can be computed recursively by

$$G(t;t_1,..,t_l) = \Sigma_{x\in T}\ G(t;x,t_l)G(x;t_1,..,t_{l-1}) =$$
$$= \Sigma_{y\in T}\ G(t;t_1,y)G(y;t_2,..,t_l) .$$

Proof: We use induction on l≥2 to show that $G(t;t_1,..,t_l)$ is well-defined. If l=2, this is (~4). For the induction step we prove the first part of the recursion formula: Fix an s∈t and return to the proof of Proposition (1.3). Define a system of representatives of $\{<p,q>\in S_2<M>;qp=s,\bar{q}=t_l\}$ by $R'=\{(p,q)\in R;\bar{q}=t_l\}$. Restricting the bijection F yields the bijection

$$\{<s_1,..,s_l>\in S_l<M>;s_l..s_1=s,\bar{s}_i=t_i \text{ for } i=1,..,l\}$$
$$\downarrow F'$$
$$\{((p,q),<r_1,..,r_{l-1}>)\in R'\times S_{l-1}<M>;r_{l-1}..r_1=p,\bar{r}_i=t_i \text{ for } i=1,..,l-1\} .$$

Since the codomain of F' is the disjoint union of the sets $\{(p,q)\}\times\{<r_1,..,r_{l-1}>\in S_{l-1}<M>;r_{l-1}..r_1=p,\bar{r}_i=t_i \text{ for } i=1,..,l-1\},(p,q)\in R'$, it follows from the induction hypothesis that

$$\#\text{domain}(F') = \Sigma_{(p,q)\in R'}\ G(\bar{p};t_1,..,t_{l-1}) =$$
$$= \Sigma_{x\in T}\ \#\{(p,q)\in R';\bar{p}=x\}\ G(x;t_1,..,t_{l-1}) .$$

As $\#\{(p,q)\in R';\bar{p}=x\} = \#\{<p,q>\in S_2<M>;qp=s,\bar{p}=x,\bar{q}=t_l\} = G(\bar{s};x,t_l)$, we end up with

$$\#\text{domain}(F') = \Sigma_{x\in T}\ G(t;x,t_l)G(x;t_1,..,t_{l-1}) .$$

The second part of the recursion formula is proved by duality: Apply the first part to $\underline{K}^{op}$,M,~ and use $G^{op}(t;t_1,..,t_l)=G(t;t_l,..,t_1)$. □

(1.10) COROLLARY: For fixed t∈T and l≥3, there are only finitely many $t_1,..,t_l\in T$ such that $G(t;t_1,..,t_l)\neq 0$. □

(1.11) COROLLARY: Let s be a morphism in M of type t.

(i) Let $l \geq 1$. Then the number of factorizations of s of length l, up to simplicial equivalence, is

$$G(t;l) = \Sigma_{t_1,\dots,t_l \in T(1)} G(t;t_1,\dots,t_l) .$$

In particular, $G(t;1)=0$ if $t \in T'$ and $G(t;1)=1$ if $t \in T(1)$.
Sometimes, it will be convenient to have

$$G(t;0)=1 \text{ if } t \in T' \quad \text{and} \quad G(t;1)=0 \text{ if } t \in T(1) .$$

(ii) The dimension of s equals the largest $l \geq 0$ such that $G(t;l) \neq 0$. Therefore, we can define the <u>dimension of a type</u> as the dimension of any representative. □

(1.12) LEMMA: Let t,t_1,t_2 be types with $G(t;t_1,t_2) \neq 0$.
Then $\dim(t) \geq \dim(t_1)+\dim(t_2)$.

Proof: When t_1 or $t_2 \in T'$, this follows from (1.8). Hence suppose that, for $i=1,2$, $d_i=\dim(t_i) \geq 1$, and fix a representative s of t. As $G(t;t_1,t_2) \neq 0$, there are morphisms $s_1,s_2 \in M$ such that $s=s_2s_1$, $\overline{s_1}=t_1$ and $\overline{s_2}=t_2$. But s_1,s_2 have factorizations of length d_1 or d_2 resp., so s has a factorization of length d_1+d_2 which implies $\dim(s) \geq d_1+d_2$. □

(1.13) LEMMA: The set of types T is partially ordered by
$t_1 \leq t_2$ if and only if there are morphisms $p,s_1,q \in M$ such that

$$\overline{s_1}=t_1 \quad \text{and} \quad \overline{ps_1q}=t_2 .$$

This order relation has the following properties:

The minimal elements of T are the types of isomorphisms.

The dimension function $\dim: T \to \mathbb{N}_0$, $t \to \dim(t)$, is strictly increasing, i.e. $t_1<t_2$ implies $\dim(t_1)<\dim(t_2)$.

All finitely generated order ideals of T are finite.

If t,t_1,t_2 are types with $G(t;t_1,t_2) \neq 0$, then $t_1,t_2 \leq t$.

Proof: Obviously, $t_1 \leq t_2$ if and only if there are types v,w such that $G(t_2;v,t_1,w) \neq 0$. Thus, if $t_1 \leq t_2$ and s_2 is an arbitrary representative of t_2, the morphisms $p,s_1,q \in M$ can be selected in such a way that $\overline{s_1}=t_1$ and $\overline{ps_1q}=s_2$. From this we infer that $\leq$ is transitive.
Clearly, $t_1 \leq t_2$ implies $\dim(t_1) \leq \dim(t_2)$. If $t_1 \leq t_2$ and $\dim(t_1)=\dim(t_2)$, then p,q must be isomorphisms, hence $t_1=t_2$ by (~1). Therefore, dim is strictly increasing and $\leq$ is antisymmetric.
Since $\mathrm{dom}(t) \leq t$ for every type t, T' is the set of minimal elements of T. By Corollary (1.10), all principal order ideals and hence all finitely generated order ideals in T are finite. □

§2. The incidence algebra

Let $\underline{K}$,M,~ be a categorical structure. As in §1, let T=M/~ be the set of types and let $G(t;t_1,t_2)$ denote the section coefficients. We now construct the incidence algebra of $\underline{K}$,M,~.

Let k be a commutative ring. In combinatorics, the ground ring usually is $\mathbb{Z}$, the ring of integers, or $\mathbb{Q}$, the field of rationals. We consider k as a topological ring with the discrete topology. In the sequel, we use some terminology from topological algebra which is explained in the appendix, §1.

With the componentwise linear operations and the product topology, k^T is a complete linear topological module over k. The topological standard basis of k^T is $(e(t);t\in T)$ where

$$e(t) = (..,0,0,1,0,0,..) \text{ , 1 at place } t \text{ .}$$

So every $f\in k^T$ has a unique representation as a convergent series

$$f = \Sigma_{t\in T} f(t)e(t) \text{ .}$$

(1.14) THEOREM: For $f,g\in k^T$, define the <u>convolution product</u> $fg\in k^T$ by

$$fg(t) = \Sigma_{t_1,t_2\in T} G(t;t_1,t_2)f(t_1)g(t_2) \quad , \quad t\in T \text{ .}$$

(In virtue of Lemma (1.7) the sum is finite). With the convolution multiplication, k^T becomes an associative topological k-algebra whose unit is the <u>delta function</u> δ given by

$$\delta(t)=1 \quad \text{if } t\in T' \quad \text{and} \quad \delta(t)=0 \quad \text{if } t\in T(1) \text{ .}$$

We call this algebra the <u>incidence algebra</u> of $\underline{K}$,M,~ and denote it by $k(T;G)$. For $t_1,t_2\in T$,

$$e(t_1)e(t_2) = \Sigma_{t\in T} G(t;t_1,t_2)e(t) \text{ .}$$

Thus the structure coefficients of $k(T;G)$ with respect to the topological basis $(e(t);t\in T)$ are the $G(t;t_1,t_2)1 \in k$.

If the section coefficients are symmetric,

i.e. if $G(t;t_1,t_2) = G(t;t_2,t_1)$ for all types t,t_1 and t_2, then the incidence algebra $k(T;G)$ is commutative.

Proof: By the definition,

$e(t_1)e(t_2)(t) = \Sigma_{v,w} G(t;v,w)e(t_1)(v)e(t_2)(w) = G(t;t_1,t_2)1$,

hence $e(t_1)e(t_2) = \Sigma_t G(t;t_1,t_2)e(t)$. Obviously, the multiplication $(f,g) \to fg$ is bilinear. It also is continuous because for a fixed type t, there are only finitely many types t_1,t_2 with $G(t;t_1,t_2)\neq 0$, and thus the map

$k^T\times k^T \to k$, $(f,g) \to fg(t)=\Sigma_{t_1,t_2}G(t;t_1,t_2)f(t_1)g(t_2)$, is continuous.

To check associativity, we calculate

$(e(t_1)e(t_2))e(t_3) = (\Sigma_x G(x;t_1,t_2)e(x))e(t_3) =$

$= \Sigma_x G(x;t_1,t_2)\Sigma_t G(t;x,t_3)e(t) = \Sigma_t(\Sigma_x G(t;x,t_3)G(x;t_1,t_2))e(t)$

and

$e(t_1)(e(t_2)e(t_3)) = e(t_1)(\Sigma_y G(y;t_2,t_3)e(y)) =$

$= \Sigma_y G(y;t_2,t_3)\Sigma_t G(t;t_1,y)e(t) = \Sigma_t(\Sigma_y G(t;t_1,y)G(y;t_2,t_3))e(t)$.

By Theorem (1.9), both expressions equal $\Sigma_t G(t;t_1,t_2,t_3)e(t)$.

Finally, we show that δ is the unit of $k(T;G)$.

For $f \in k^T$ and $t \in T$, Lemma (1.8) implies that

$\delta f(t) = \Sigma_{t_1,t_2} G(t;t_1,t_2)\delta(t_1)f(t_2) = f(t)$ and

$f\delta(t) = \Sigma_{t_1,t_2} G(t;t_1,t_2)f(t_1)\delta(t_2) = f(t)$. □

(1.15) PROPOSITION: Let $l \geqq 3$.

(i) The convolution product of $f_1,..,f_l \in k(T;G)$ can be computed by

$$f_1..f_l(t) = \Sigma_{t_1,..,t_l \in T} G(t;t_1,..,t_l)f_1(t_1)..f_l(t_l) \quad , \quad t \in T \ .$$

(ii) For types $t_1,..,t_l$, $e(t_1)..e(t_l) = \Sigma_{t \in T} G(t;t_1,..,t_l)e(t)$.

Proof: (i) By induction on l and the recursion formula (1.9) we have

$f_1..f_l(t) = (f_1..f_{l-1})f_l(t) = \Sigma_{x,t_l} G(t;x,t_l)f_1..f_{l-1}(x)f_l(t_l) =$

$= \Sigma_{x,t_l} G(t;x,t\ )\Sigma_{t_1,..,t_{l-1}} G(x;t_1,..,t_{l-1})f_1(t_1)..f_{l-1}(t_{l-1})f_l(t_l) =$

$= \Sigma_{t_1,..,t_l}(\Sigma_x G(t;x,t_l)G(x;t_1,..,t_{l-1}))f_1(t_1)..f_l(t_l) =$

$= \Sigma_{t_1,..,t_l} G(t;t_1,..,t_l)f_1(t_1)..f_l(t_l)$.

(ii) Applying (i) yields $e(t_1)..e(t_l)(t) = G(t;t_1,..,t_l)1$. □

(1.16) THEOREM:

(i) $$V = \{f \in k^T; f(t)=0 \text{ if } t \in T(1)\}$$

is a closed unitary subalgebra of $k(T;G)$ isomorphic to the product algebra $k^{T'}$ by $V \to k^{T'}$, $f \to (f(t))_{t \in T'}$.

(ii) The dimension function on T induces the filtration of $k(T;G)$

$$k(T;G)=I(0) \supset I(1) \supset I(2) \supset \ldots$$

by the closed two-sided ideals

$$I(d) = \{f \in k^T; f(t)=0 \text{ if } \dim(t) \leq d-1\} \ .$$

This filtration converges to $\{0\}$, i.e. for any neighborhood N of 0 there is a number d such that $I(d) \subset N$. For all $d_1,d_2 \geq 0$,

$$I(d_1)I(d_2) \subset I(d_1+d_2) \ .$$

In particular, the elements of $I(1)$ are topologically nilpotent.

(iii) $k(T;G)$ is the topological direct sum of V and $I(1)$.

Proof: (i) For $f,g\in V$ and $t\in T$,
$fg(t) = \Sigma_{t_1,t_2} G(t;t_1,t_2)f(t_1)g(t_2) = f(t)g(t)$ by Lemma (1.8). Hence V is a subalgebra of $k(T;G)$ and the map $V \to k^{T'}$, $f \to (f(t))_{t\in T'}$, is an isomorphism.
(ii) Obviously, the $I(d)$ are closed submodules of k^T with the topological bases $(e(t);t\in T,\dim(t)\geq d)$. If t_1,t_2 are types with $\dim(t_1)\geq d_1$ and $\dim(t_2)\geq d_2$, then Lemma (1.12) implies that
$e(t_1)e(t_2) = \Sigma\, G(t;t_1,t_2)e(t)$ where the summation is taken over all types t with $\dim(t)\geq\dim(t_1)+\dim(t_2)$. Thus $e(t_1)e(t_2)\in I(d_1+d_2)$ which establishes $I(d_1)I(d_2)\subset I(d_1+d_2)$. In particular, since $I(0)=k(T;G)$, the $I(d)$ are two-sided ideals of $k(T;G)$. It remains to show that the filtration $k(T;G) \supset I(1) \supset I(2) \supset ..$ converges to $\{0\}$. But every neighborhood N of 0 contains a neighborhood of the form $\{f\in k^T;f(t)=0 \text{ for all } t\in S\}$ where S is a finite subset of T. Choosing d such that $d\geq\dim(t)+1$ for all $t\in S$, we have $I(d)\subset N$.
(iii) $(e(t);t\in T')$ is a topological basis of V, $(e(t);t\in T(1))$ is a topological basis of $T(1)$ and T is the disjoint union of T' and $T(1)$. □

(1.17) PROPOSITION: Let J be an order ideal of T. Then $J^{\perp} = \{f\in k^T;f(t)=0 \text{ for all } t\in J\}$ is a closed two-sided ideal of $k(T;G)$. If J is finitely generated, $J^{\perp}$ is open.

Proof: Clearly $J^{\perp}$ is a closed submodule of k^T. Given $f\in J^{\perp}$, $g\in k^T$ and $t\in J$, it follows from Lemma (1.13) that
$fg(t) = \Sigma_{t_1,t_2\leq t} G(t;t_1,t_2)f(t_1)g(t_2) = 0$, and similarly that $gf(t)=0$. Thus $J^{\perp}$ is a two-sided ideal of $k(T;G)$. If J is finitely generated, then it is finite by (1.13). □

(1.18) COROLLARY: The set $\{J^{\perp};\ J$ a finitely generated order ideal of $T\}$ is a basis of neighborhoods of 0 consisting of two-sided ideals of $k(T;G)$. □

(1.19) LEMMA: Define closed submodules of k^T by

$$V(d) = \{f\in k^T;f(t)=0 \text{ if } \dim(t)\neq 0\} \quad , \quad d=0,1,2,.. \; .$$

In particular, $V(0)=V$. Then $I(d) = V(d)\hat{\oplus}I(d+1)$ for all d, and

$$k^T = \Pi_{d\geq 0}\, V(d) \quad .$$

Proof: The $V(d)$ have the topological bases $(e(t);t\in T,\dim(t)=d)$ whereas the $I(d)$ have the topological bases $(e(t);t\in T,\dim(t)\geq d)$. □

(1.20) THEOREM: The topological algebra $k(T;G)$, together with the subalgebra V and the filtration

$$k(T;G) \supset I(1) \supset I(2) \supset \ldots \quad ,$$

is an <u>abstract incidence algebra</u> over k (Abstract incidence algebras are defined and studied in the appendix, §2, as an appropriate frame for a "generating function calculus").

$k(T;G)$ is <u>graded</u> by the sequence $(V(d))_{d\geq 0}$, i.e.

$$V(d_1)V(d_2) \subset V(d_1+d_2) \qquad \text{for all } d_1,d_2 \quad ,$$

if one of the following equivalent conditions is satisfied:

(i) For all composable morphisms s_1,s_2 in M,

$$\dim(s_2s_1) = \dim(s_1)+\dim(s_2) \quad .$$

(ii) For every morphism s in M, all maximal factorizations of s have equal length.

If the characteristic of k is 0, then the conditions (i),(ii) are also necessary.

Proof: Collecting the results from (1.16),(1.18) and (1.19), we see that $k(T;G)$ is an AIA as defined in (A.2). It remains to demonstrate when $k(T;G)$ is graded by $(V(d))_{d\geq 0}$. Recall that $V(d)$ has the topological basis $(e(t);t\in T,\dim(t)=d)$ and that $e(t_1)e(t_2) = \Sigma_t G(t;t_1,t_2)e(t)$. Hence $V(d_1)V(d_2)\subset V(d_1+d_2)$ for all d_1,d_2 if $\dim(t)=\dim(t_1)+\dim(t_2)$ for all types t,t_1,t_2 with $G(t;t_1,t_2)\neq 0$. But this is equivalent to condition (i) or (ii). If k has characteristic 0, we can argue conversely. □

(1.21) COROLLARY: An element f is invertible in $k(T;G)$ if and only if $f(t)$ is invertible in k for all $t\in T'$.

Proof: Write $f=g+h$ according to the decomposition $k(T;G) = V \hat{\oplus} I(1)$. By Proposition (A.5) from the appendix, f is invertible in $k(T;G)$ if and only if g is invertible in V. Since V is isomorphic to the product algebra $k^{T'}$ by $g \to (g(t))_{t\in T'}$, and since $g(t)=f(t)$ for all $t\in T'$, the conclusion follows. □

For an arbitrary ring B, we denote the group of invertible elements of B by $E(B)$.

With the topology induced from $k(T;G)$, $E(k(T;G))$ becomes a Hausdorff topological group. Now, consider the decomposition $k(T;G) = V \hat{\oplus} I(1)$ from Theorem (1.16). Obviously, $E(V)$ is a closed subgroup of $E(k(T;G))$ isomorphic to the product group $E(k)^{T'}$, and

$\delta+I(1) = \{\delta+h;h\in I(1)\}$ is a closed normal subgroup of $E(k(T;G))$.

(1.22) PROPOSITION: The group $E(k(T;G))$ is the semidirect product of the subgroups $E(V)$ and $\delta+I(1)$. The action of $E(V)$ on $\delta+I(1)$ by inner automorphisms is given by

$$bcb^{-1}(t) = \frac{b(\mathrm{dom}(t))}{b(\mathrm{cod}(t))}\, c(t) \quad , \quad t\in T \quad ,$$

where $b\in E(V)$ and $c\in\delta+I(1)$.

Proof: The first statement is (A.15). To compute bcb^{-1}, write $b = \Sigma_{v\in T'} b(v)e(v)$, $c = \delta+\Sigma_{t\in T(1)} c(t)e(t)$ and $b^{-1} = \Sigma_{w\in T'} b(w)^{-1}e(w)$.
Then, by Lemma (1.8),

$bc = b+\Sigma_{t\in T(1)} b(\mathrm{dom}(t))c(t)e(t)$ and

$bcb^{-1} = \delta+\Sigma_{t\in T(1)} b(\mathrm{dom}(t))c(t)b(\mathrm{cod}(t))^{-1}e(t)$. □

For any commutative k-algebra R (endowed with the discrete topology), we can construct the incidence algebra $R(T;G)$ as for the ground ring k. Algebraically, this corresponds to the base ring extension $k(T;G)\hat{\otimes}_k R$ where $\hat{\otimes}_k$ denotes the completed tensor product over k (appendix, §1). In an obvious way, $k(T;G)\hat{\otimes}_k R$ can be identified with $R(T;G)$.
From Theorem (A.24) we derive the main result on the structure of incidence algebras.

(1.23) THEOREM: Suppose that k is a field.
Then the affine group

$$\Omega\colon \underline{Al}_k \to \underline{Gr} \ , \ R \to E(R(T;G)) \ ,$$

of invertible elements in the incidence algebra
is the semidirect product of the closed <u>diagonalizable</u> subgroup

$$K\colon \underline{Al}_k \to \underline{Gr} \ , \ R \to E(V\hat{\otimes}_k R) \ ,$$

and the closed <u>unipotent</u> normal subgroup

$$\Lambda\colon \underline{Al}_k \to \underline{Gr} \ , \ R \to \delta+I(1)\hat{\otimes}_k R \ ,$$

and hence is <u>triangulable</u>. □

In the following definition, we introduce special elements of the incidence algebra which are familiar in the combinatorial literature (compare e.g. [Ro,64],[Ai,79],[CF,69] or [CLL,80]).

(1.24) DEFINITION: Let k be the ring of integers.
(i) We call $\zeta \in k^T$, $\zeta(t)=1$ for all $t\in T$, the <u>zeta function</u> in $k(T;G)$ and its inverse $\mu = \zeta^{-1}$ the <u>Möbius function</u> in $k(T;G)$
(This terminology is justified by Example (1.58) where μ is the classical Möbius function from number theory and ζ corresponds to the Riemann zeta function).

In the literature, the use of $\mu=\zeta^{-1}$ is called Möbius inversion . Applying Proposition (A.14) to $\zeta = \delta+\Sigma_{t\in T(1)}e(t) \in \delta+I(1)$, we find that $\mu = \delta+\Sigma_{t\in T(1)}(\Sigma_{l=1}^{\infty}(-1)^{l}G(t;l))e(t)$, thus

$$\mu(t)=1 \quad \text{if } t\in T' \quad \text{and} \quad \mu(t) = \Sigma_{l=1}^{\dim(t)}(-1)^{l}G(t;l) \quad \text{if } t\in T(1) \ .$$

Recall from Corollary (1.11) that, given a morphism s in M of type t, G(t;l) counts the factorizations of s of length l up to simplicial equivalence. By Proposition (1.15),

$$(\zeta-\delta)^{l}(t) = G(t;l) \quad .$$

Since $\zeta-\delta\in I(1)$ is topologically nilpotent and $\Sigma_{l=0}^{\infty}(\zeta-\delta)^{l} = (2\delta-\zeta)^{-1}$, the total number of factorizations of s up to simplicial equivalence is

$$\Sigma_{l=0}^{\dim(t)} G(t;l) = (2\delta-\zeta)^{-1}(t) \quad .$$

(ii) Let $A=k[D]$ be the polynomial algebra in the indeterminate D, and define $D^{\dim} \in A(T;G)$ by $D^{\dim}(t) = D^{\dim(t)}$ where dim denotes the dimension function for types. Then $\chi = \mu D^{\dim} \in A(T;G)$ is called the characteristic function in A(T;G).

Of course, for any commutative ring R, one can consider zeta, Möbius and characteristic function in R(T;G) or R[D](T;G) by extending the base ring from k to R.

Möbius inversion has been studied by P.Cartier, D.Foata et alii on monoids with the finite factorization property ([CF,69],[Des,80]), and by G.C.Rota et alii on locally finite partially ordered sets ([Ro,64],[Cr,66],[Cr,68],[Li,69],[DRS,72],[St,74],...). We briefly show how their constructions are special cases of the general theory presented here.

(1.25) Monoids with the finite factorization property:

Let N be a monoid which enjoys the "finite factorization property":

For every $x\in N$, there are only finitely many ways to write x as a product $x = x_1..x_l$ where $l\geq 1$ and no x_i is the unit 1 of N.

For instance, the monoids L(Z;C) introduced by P.Cartier and D.Foata in [CF,69],p.9, have the ffp (Given a non-empty set Z and a relation C on Z, L(Z;C) is the monoid freely generated by Z subject to the "commutation relations" C. It can be constructed as a factor of the free monoid of words over the alphabet Z). Another example is the "plactic monoid" studied by A.Lascoux and M.P.Schützenberger in [LS,78].

Let $\underline{K}$ be the category with a single object whose morphisms are the elements of N. In $\underline{K}$ the sole isomorphism is the identity 1 because N has no invertible elements except the unit 1 (owing to the ffp).
Set M=N . As simplicial equivalence reduces to identity, condition (M3) follows from the ffp of N. If N=L(Z;C), the factorizations of a morphism [w]∈L(Z;C) are just the "decompositions" of [w] defined in [CF,69],p.18, and the dimension of [w] is the length l(w) of the word w. For equivalence choose isomorphy in Mor($\underline{K}$), i.e. the identity relation. Then T=N by a trivial identification.

Since G(x;y,z)=1 if yz=x and G(x;y,z)=0 otherwise, the convolution product of $f,g\in k^N$ is given by

$$fg(x) = \Sigma_{yz=x} f(y)g(z) \quad , \quad x\in N \quad .$$

Hence the incidence algebra of $\underline{K}$,M,~ is the <u>large monoid algebra</u> of N. In particular, the topological word algebra and the power series algebra are incidence algebras.

In [CF,69],p.21, Cartier and Foata determine the Möbius function μ in $\mathbb{Z}(L(Z;C);G)$:

$$\mu([w]) = (-1)^{l(w)} \quad \text{if all letters in w are different and commute,}$$

and

$$\mu([w]) = 0 \quad \text{otherwise.}$$

The Möbius function of the plactic monoid has been computed by J.Désarménien using a non-commutative analogue of the Littlewood--Richardson multiplication rule for Schur functions ([Des,80]).

(1.26) <u>Locally finite partially ordered sets:</u>

Let P be a locally finite partially ordered set (i.e. P is partially ordered - a poset - , and all intervals of P are finite).

From P we construct a category $\underline{K}$: The objects in $\underline{K}$ are the elements of P, and the morphisms in $\underline{K}$ are the pairs $(x,y)\in P\times P$ where $x\leq y$. The domain of (x,y) is x, the codomain is y. The composition of (x,y) and (y,z) is (x,z), provided that $x\leq y\leq z$. In $\underline{K}$ the sole isomorphisms are the identities (x,x), x∈P.
Set $M=\{(x,y)\in P\times P; x\leq y\}$. Then the factorizations of a morphism (x,y) correspond to the chains in P from x to y, and the dimension of (x,y) is the maximal length of a chain from x to y. Since simplicial equivalence reduces to identity, condition (M3) follows from the local finiteness of P. For equivalence we take isomorphy in Mor($\underline{K}$), i.e. the identity relation, and identify $T=\{(x,y)\in P\times P; x\leq y\}$.

As $G((x,y);(x,z),(z,y))=1$ for all z with $x \leq z \leq y$ and $G((x,y);t_1,t_2)=0$ otherwise, the convolution product of $f,g \in k^T$ is given by

$$fg(x,y) = \Sigma_{x \leq z \leq y} f(x,z)g(z,y) \quad , \quad (x,y) \in T \quad .$$

In the literature, the algebra $k(T;G)$ is called the <u>incidence algebra of the poset P</u> and is denoted by $I(P,k)$ ([Ro,64],[DRS,72]). For a field K, R.P.Stanley has shown that, given two locally finite posets P and Q, the incidence algebras $I(P,K)$,$I(Q,K)$ are isomorphic as rings if and only if P,Q are isomorphic as posets ([St,70] or [DRS,72]). If P is finite, $I(P,k)$ can be embedded in the algebra of upper triangular matrices of order $n=\#P$ with entries in k ([Ai,79],p.140). From this we derive a faithful linear representation of the affine algebraic group $\Omega:\underline{Al}_k \to \underline{Gr}$, $R \to E(I(P,R))$, of invertible elements of the incidence algebra by upper triangular matrices of order n. By Theorem (1.23) we know a priori that such a representation must exist if k is a field.

(1.27) Finally, choosing an arbitrary equivalence relation ~ on M with the properties (~1)-(~4), we obtain the "<u>reduced incidence algebras</u>" of [DRS,72] (see also [Ai,79] and [JR,79]). In the last article, instead of topologically k-free algebras, dually, discrete k-free coalgebras are considered.

An example similar to the Boolean coalgebra of [JR,79],pp.100, but not mentioned there, is the <u>Wang algebra</u> from graph theory which can be used for the generation of trees in electrical networks ([Ch,76]).

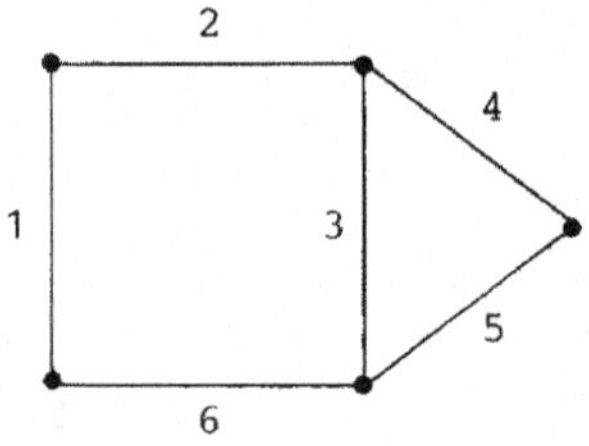

Let H be an undirected finite graph with edge set E. We represent a subgraph of H containing no isolated nodes by its edge set. Then the set P of all subgraphs of H containing no isolated nodes is the power set Pot(E). With the inclusion relation, $P=\mathrm{Pot}(E)$ is a finite poset. Define the equivalence $\sim$ on $M=\{(x,y) \in P \times P; x \subset y\}$ by

$(x_1,y_1) \sim (x_2,y_2)$ if and only if $y_1-x_1=y_2-x_2$.

Identifying a type $\overline{(x,y)}$ with the subset $y-x$ of E, we have $T=P$ and $G(x;y,z)=1$ if x is the disjoint union of y and z, and $G(x;y,z)=0$ else.

Let $\mathbb{F}=\{0,1\}$ be the field with two elements. Then the convolution product of $f,g\in\mathbb{F}^P$ is given by

$$fg(x) = \Sigma_{y\dot{\cup}z=x} f(y)g(z) \quad , \quad x\in P \quad .$$

The incidence algebra $\mathbb{F}(P;G)$ is commutative and has the finite $\mathbb{F}$-basis $(e(x);x\in P)$. For $y,z\in P$,

$$e(y)e(z)=e(y\cup z) \text{ if } y,z \text{ are disjoint,} \quad \text{and} \quad e(y)e(z)=0 \text{ otherwise.}$$

But an element of $\mathbb{F}^P$ can be interpreted as a set of subgraphs of H containing no isolated nodes, by considering $f^{-1}(1) \subset P$. Carrying the $\mathbb{F}$-algebra structure from $\mathbb{F}(P;G)$ to Pot(P), we obtain the Wang algebra Pot(P) where addition is called ring sum and multiplication the Wang product. Obviously, ring sum is symmetric difference.
For instance, consider the connected graph H depicted above. Here $E=\{1,2,3,4,5,6\}$, and

$$C_1=\{2,6\} \ , \ C_2=\{2,3,5\} \ , \ C_3=\{3,4,6\} \text{ and } C_4=\{1,2\}$$

are independent cuts of H. Then a general theorem in [Ch,76] asserts that the Wang product of the sets $\{\{2\},\{6\}\}$, $\{\{2\},\{3\},\{5\}\}$, $\{\{3\},\{4\},\{6\}\}$ and $\{\{1\},\{2\}\}$ gives the set of all spanning trees of H. Indeed, for $f_1=e(\{2\})+e(\{6\})$, $f_2=e(\{2\})+e(\{3\})+e(\{5\})$, $f_3=e(\{3\})+e(\{4\})+e(\{6\})$ and $f_4=e(\{1\})+e(\{2\})$, we have

$$\begin{aligned} f_1f_2f_3f_4 = {} & e(\{1,2,3,4\})+e(\{1,2,3,5\})+e(\{1,2,4,5\})+e(\{1,2,4,6\})+ \\ & +e(\{1,2,5,6\})+e(\{1,3,4,6\})+e(\{1,3,5,6\})+e(\{1,4,5,6\})+ \\ & +e(\{2,3,4,6\})+e(\{2,3,5,6\})+e(\{2,4,5,6\}) \ , \end{aligned}$$

and the spanning trees of H are
$\{1,2,3,4\}$, $\{1,2,3,5\}$, $\{1,2,4,5\}$, ...

(1.28) **REMARK:** Let $\underline{K},M,\sim$ be a categorical structure with the set of types $T=\{\bar{s};s\in M\}$ and the section coefficients $G(t;t_1,t_2)$. For the isomorphism relation $\simeq$ on M we denote the set of isomorphism types by $I=\{\hat{s};s\in M\}$. Let $\Gamma(\alpha;\alpha_1,\alpha_2)$ be the section coefficients of the categorical structure $\underline{K},M,\simeq$. Because of ($\sim$1) we have a map can: $I \to T$, $\hat{s} \to \bar{s}$, and hence a continuous k-module homomorphism

$$\text{em: } k^T \to k^I, \ f \to f\circ\text{can} \ .$$

Since can is onto, em is one-to-one. By ($\sim$2), em(δ) is the delta function in $k(I;\Gamma)$. Finally, recalling ($\sim$4), we see that $G(\text{can}(\alpha);t_1,t_2) = \Sigma\ \Gamma(\alpha;\alpha_1,\alpha_2)$ where the sum runs over all $\alpha_1,\alpha_2\in I$ with $\text{can}(\alpha_1)=t_1$ and $\text{can}(\alpha_2)=t_2$. Hence, by a straight-forward calculation, $\text{em}(fg) = \text{em}(f)\text{em}(g)$ for all $f,g\in k(T;G)$.
We conclude that $k(T;G)$ is isomorphic to a closed subalgebra of $k(I;\Gamma)$. Under this isomorphism, the zeta and the Möbius function are preserved.

(1.29) Products of categorical structures: Let $\underline{K}_1, M_1, \sim_1$ and $\underline{K}_2, M_2, \sim_2$ be two categorical structures. In the product category $\underline{K} = \underline{K}_1 \times \underline{K}_2$, consider the class of morphisms $M = M_1 \times M_2$ and the equivalence relation on M $(s_1, s_2) \sim (r_1, r_2)$ if and only if $s_1 \sim_1 r_1$ and $s_2 \sim_2 r_2$. It is easy to verify that $\underline{K}, M, \sim$ is a new categorical structure. We identify types $\overline{(s_1, s_2)} \in T = M/\sim$ with pairs of types $(\overline{s_1}, \overline{s_2}) \in T_1 \times T_2$. Then the section coefficients are

$G((t_1, t_2); (v_1, v_2), (w_1, w_2)) = G(t_1; v_1, w_1) G(t_2; v_2, w_2)$

where $t_1, v_1, w_1 \in T_1$ and $t_2, v_2, w_2 \in T_2$. Hence the incidence algebra of $\underline{K}, M, \sim$ is isomorphic to the topological tensor product of the incidence algebras of $\underline{K}_1, M_1, \sim_1$ and $\underline{K}_2, M_2, \sim_2$ by

$$k(T;G) \to k(T_1;G_1) \hat{\otimes}_k k(T_2;G_2) \ , \ e(t_1, t_2) \to e(t_1) \hat{\otimes} e(t_2) \ .$$

(1.30) Substructures: Let $\underline{K}, M, \sim$ be a categorical structure. Let M_o be a subclass of M which in addition to (M1) has the following property:

M_o is saturated with respect to $\sim$,

i.e. for any $s_1 \in M_o$ and $s_2 \in M$, $s_1 \sim s_2$ implies that $s_2 \in M_o$.

Then $\underline{K}, M_o$ and the restricted relation $\sim$ also constitute a categorical structure, the types form a subset $T_o \subset T$ and the section coefficients are just the $G(t; t_1, t_2)$ where $t, t_1, t_2 \in T_o$. The incidence algebra $k(T_o; G_o)$ is isomorphic to a closed subalgebra of $k(T;G)$ by

$$k(T_o; G_o) \to k(T;G) \ , \ f_o \to f \text{ where } f(t) = f_o(t) \text{ if } t \in T_o \text{ and } f(t) = 0 \text{ else.}$$

§3. Subobjects and quotient objects

In this section we show how the partially ordered sets of subobjects or quotient objects in a "combinatorial category" can be studied by the incidence algebra of a suitable categorical structure. Our approach is inspired by the construction of the Hall algebra from the subgroup lattices of abelian p-groups ([Ha,59],[Ma,79]). Examples are given in §5.

Let $\underline{K}$ be a skeletal-small category. We distinguish two cases.

First case:

Let M be a class of monomorphisms in $\underline{K}$ with the following property:

(S1) M contains all isomorphisms in $\underline{K}$ and is closed under composition.

For any object Y in $\underline{K}$, let $M(\rightarrow Y)$ denote the class of all monomorphisms in M with codomain Y. We define a preorder on $M(\rightarrow Y)$ by putting $s_1 \leq s_2$ if and only if there is a morphism s in $\underline{K}$ such that $s_1 = s_2 s$. If s exists, it is a uniquely determined monomorphism and is denoted by $s_2 \backslash s_1$. We assume:

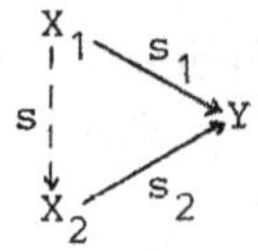

(S2) For all $s_1, s_2 \in M$ with $s_1 \leq s_2$, also $s = s_2 \backslash s_1 \in M$.

Obviously, $s_1 \leq s_2$ and $s_2 \leq s_1$ if and only if $s_2 \backslash s_1$ is an isomorphism. The classes in $M(\rightarrow Y)$ with respect to this equivalence relation are called the <u>subobjects</u> of Y in M and will be denoted by [..]. The set of subobjects of Y in M

$$\mathrm{Sub}_M(Y) = \{[r]; r \in M(\rightarrow Y)\}$$

is partially ordered by the relation induced from $\leq$, and has the greatest element $[id_Y]$. On the other hand we denote the poset of all subobjects of Y by Sub(Y).

It is often convenient to identify the class $[r: X \rightarrow Y] \in \mathrm{Sub}(Y)$ with a distinguished representative $Z \rightarrow Y$, and to call Z a subobject of Y instead of $[Z \rightarrow Y]$. For instance, in the categories of sets, groups or modules, the distinguished representative is the inclusion map, and the subobjects of Y correspond to the subsets, subgroups or submodules of Y, respectively.

The last condition for M is

(S3) For every object Y in $\underline{K}$, the poset $\mathrm{Sub}_M(Y)$ is locally finite (i.e. all intervals are finite).

(1.31) LEMMA: Let l be an integer ≥ 2 and let $s: X \rightarrow Y$ be a monomorphism in M. Then we have inverse bijections

$$\{\langle s_1, \ldots, s_l \rangle \in S_l\langle M \rangle; s_l \ldots s_1 = s\}$$

$$\downarrow \qquad \uparrow$$

$$\{([r_1], \ldots, [r_{l-1}]) \in \mathrm{Sub}_M(Y)^{l-1}; [s] \leq [r_1] \leq \ldots \leq [r_{l-1}] \leq [id_Y]\},$$

$$\langle s_1, \ldots, s_l \rangle \longmapsto ([s_l \ldots s_2], \ldots, [s_l s_{l-1}], [s_l])$$

$$([r_1], \ldots, [r_{l-1}]) \longmapsto \langle r_1 \backslash s, r_2 \backslash r_1, \ldots, r_{l-1} \backslash r_{l-2}, r_{l-1} \rangle .$$

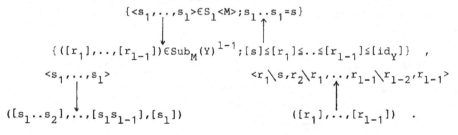

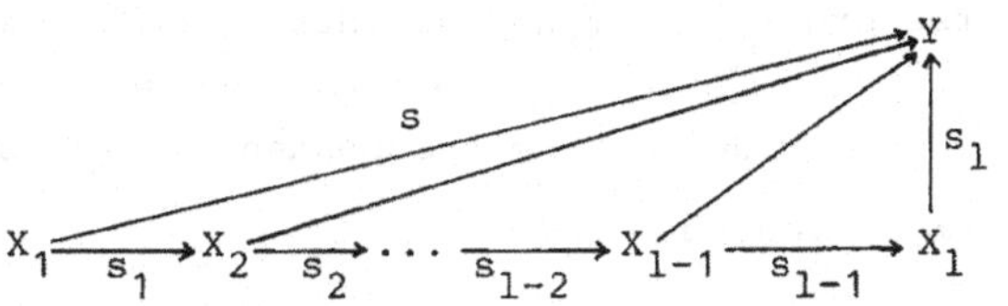

Proof: It is not difficult to check that both maps are defined independently of the representative and inverse to each other. □

(1.32) COROLLARY: Let $s:X \to Y$ be a monomorphism in M. Then the dimension of s equals the maximal length of a chain in $\mathrm{Sub}_M(Y)$ from [s] to $[\mathrm{id}_Y]$. □

We conclude that M satisfies the conditions (M1)-(M3): (M1) is (S1), (M2) holds because M is a class of monomorphisms, and (M3) follows from (S3) by Proposition (1.3).
Now let ~ be an equivalence relation on M with the properties (~1)-(~3).

(1.33) COROLLARY: Condition (~4) holds if and only if, for all types t, t_1 and t_2, the number

$$\#\{[r]\in \mathrm{Sub}_M(Y);[s]\leq[r]\leq[\mathrm{id}_Y],\overline{r\backslash s}=t_1,\bar{r}=t_2\}$$

is independent of the choice of the representative set. Then this number is the section coefficient $G(t;t_1,t_2)$. □

In what follows we also assume (~4). Hence $\underline{K}$,M,~ is a categorical structure.

Let Y be an object in $\underline{K}$. Given an ascending sequence in $\mathrm{Sub}_M(Y)$

$$[r_0]\leq[r_1]\leq\ldots\leq[r_{l-1}]\leq[r_l] \quad , \quad l\geq 1 \quad ,$$

we call

$$(\overline{r_1\backslash r_0},\overline{r_2\backslash r_1},\ldots,\overline{r_l\backslash r_{l-1}}) \in T^l$$

the <u>associated type sequence</u> . The ascending sequence $[r_0]\leq\ldots\leq[r_l]$ is a chain of length l (i.e. all subobjects in the sequence are different) if and only if all associated types are contained in T(1).
By the following Lemma it suffices to consider ascending sequences with end point $[\mathrm{id}_Y]$.

(1.34) LEMMA: Let $s_2:X_2\to Y$ be a monomorphism in M. Then the maps

$$\begin{array}{ccc} \{[s_1]\in \mathrm{Sub}_M(Y);[s_1]\leq[s_2]\} & \rightleftarrows & \mathrm{Sub}_M(X_2) \\ [s_1] & \longrightarrow & [s_2\backslash s_1] \\ [s_2 s] & \longleftarrow & [s] \end{array}$$

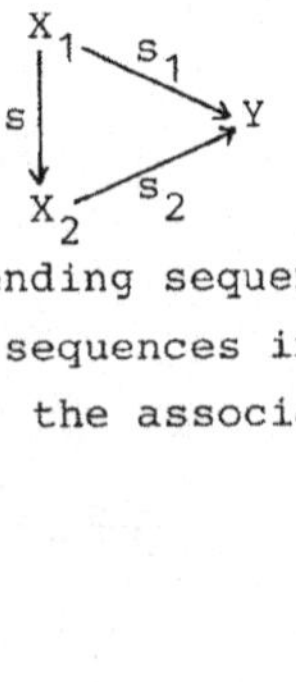

are mutually inverse order isomorphisms. Thus ascending sequences in $\mathrm{Sub}_M(Y)$ from $[s_1]$ to $[s_2]$ correspond to ascending sequences in $\mathrm{Sub}_M(Y_2)$ from $[s_2\backslash s_1]$ to $[\mathrm{id}_{X_2}]$. Under this correspondence, the associated type sequence is preserved. □

From Lemma (1.31) we derive a new interpretation of the higher section coefficients.

(1.35) PROPOSITION: Let $[s_1],[s_2]$ be subobjects of Y in M such that $[s_1]\leq[s_2]$, and let t be the type of $s_2\setminus s_1$.
Then the number of ascending sequences in $\mathrm{Sub}_M(Y)$ from $[s_1]$ to $[s_2]$ with associated type sequence $(t_1,..,t_l)\in T^l$, $l\geq 1$, is $G(t;t_1,..,t_l)$. For every $l\geq 0$, there are exactly $G(t;l)$ chains of length l in $\mathrm{Sub}_M(Y)$ from $[s_1]$ to $[s_2]$. In particular, the maximal length of a chain from $[s_1]$ to $[s_2]$ equals dim(t). □

By Theorem (1.20), this implies the following result.

(1.36) COROLLARY: Suppose that the ground ring k has characteristic 0. Then the incidence algebra k(T;G) is graded by the sequence $(V(d))_{d\geq 0}$ if and only if all posets $\mathrm{Sub}_M(Y)$ satisfy the (Jordan-Dedekind) chain condition (that is, all maximal chains between two points have equal length). □

Finally we establish the connection between the incidence algebra k(T;G) and the incidence algebras of the posets $\mathrm{Sub}_M(Y)$, Y an object in K.

(1.37) THEOREM: For any commutative ring k, the map

$$l_Y\colon k(T;G) \to I(\mathrm{Sub}_M(Y),k) \quad , \quad f \to l_Y(f) \quad ,$$

defined by

$$l_Y(f)([s_1],[s_2])=f(t) \quad \text{where t is the type of } s_2\setminus s_1 \quad ,$$

is a continuous k-algebra homomorphism.

In particular, let k be the ring of integers. Then l_Y preserves the zeta, Möbius and dimension function and, having extended the base ring to k[D], also the characteristic function. For instance, if $\zeta,\mu,\dim,\chi$ denote the zeta, Möbius, dimension and characteristic function in k(T;G) or k[D](T;G) resp., we have

$$\zeta^2(t) = \#\{[r]\in\mathrm{Sub}_M(Y);[s_1]\leq[r]\leq[s_2]\} \quad ,$$

$$\mu(t) = \mu_{\mathrm{Sub}_M(Y)}([s_1],[s_2]) \quad \text{and}$$

$$(\zeta-\delta)^l(t) = \text{number of chains of length } l \text{ in } \mathrm{Sub}_M(Y) \text{ from } [s_1] \text{ to } [s_2] \quad .$$

Furthermore, if $\mathrm{Sub}_M(Y)$ satisfies the chain condition, then dim(t) is the rank of the interval $\{[r]\in\mathrm{Sub}_M(Y);[s_1]\leq[r]\leq[s_2]\}$, $\chi(t)$ is its characteristic polynomial ([Ro,64],p.343), and

$$D^{\dim}\zeta(t) = \Sigma_{i=0}^{\dim(t)} W(i)D^{i}$$

where the $W(i)$ are the Whitney numbers ([GK,78],p.24) which count the elements of rank i respectively.
(More precisely, the $W(i)$ are the Whitney numbers of the second kind whereas the coefficients of the characteristic polynomial are the Whitney numbers of the first kind. In [Ai,79] they are called level numbers of the first or second kind.)

Proof: Obviously, l_Y is k-linear and continuous. Now consider two subobjects $[s_1]\leq[s_2]$ of Y and denote the type of $s_2\backslash s_1$ by t. Since $[s_1]=[s_2]$ if and only if $t\in T'$, $l_Y(\delta)$ is the delta function in $I(\mathrm{Sub}_M(Y),k)$. To show that l_Y preserves the convolution product, let $f,g\in k(T;G)$. Then $l_Y(f)l_Y(g)([s_1],[s_2]) =$

$= \Sigma_{[s_1]\leq[r]\leq[s_2]} f(\overline{r\backslash s_1})g(\overline{s_2\backslash r}) = \Sigma_{t_1,t_2\in T} G(t;t_1,t_2)f(t_1)g(t_2) =$

$= fg(t) = l_Y(fg)([s_1],[s_2])$ by Proposition (3.15). □

Second case:

Let M be a class of epimorphisms in $\underline{K}$ with the following property:

(Q1) M contains all isomorphisms in $\underline{K}$ and is closed under composition.

For any object X in $\underline{K}$, let $M(X\rightarrow)$ denote the class of all epimorphisms in M with domain X. We define a preorder on $M(X\rightarrow)$ by putting $s_1\leq s_2$ if and only if there is a morphism s in $\underline{K}$ such that $s_2=ss_1$. If s exists, it is a uniquely determined epimorphism and is denoted by s_2/s_1. We assume:

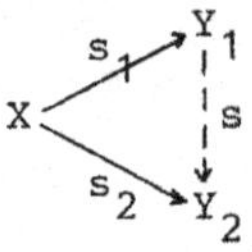

(Q2) For all $s_1,s_2\in M$ with $s_1\leq s_2$, also $s=s_2/s_1\in M$.

Obviously, $s_1\leq s_2$ and $s_2\leq s_1$ if and only if s_2/s_1 is an isomorphism. The classes in $M(X\rightarrow)$ with respect to this equivalence relation are called the quotient objects of X in M and will be denoted by [..]. The set of quotient objects of X in M

$$\mathrm{Qu}_M(X) = \{[r];r\in M(X\rightarrow)\}$$

is partially ordered by the relation induced from $\leq$, and has the least element $[id_X]$. On the other hand we denote the poset of all quotient objects of X by $\mathrm{Qu}(X)$.
It is often convenient to identify the class $[r:X\rightarrow Y] \in \mathrm{Qu}(X)$ with a distinguished representative $X\rightarrow Z$, and to call Z a quotient object of X

instead of $[X \to Z]$. For instance, in the categories of sets, groups or modules, the distinguished representative is the quotient map, and the quotient objects of X correspond to the partitions, quotient groups or quotient modules of X, respectively.
The last condition for M is

(Q3) For every object X in $\underline{K}$, the poset $Qu_M(X)$ is locally finite (i.e. all intervals are finite).

In the sequel, we state the analogous results to the first case which can be proved by duality. Observe that dualization yields both the opposite order relation and the opposite algebras.

(1.38) LEMMA: Let l be an integer ≥ 2 and let $s: X \to Y$ be an epimorphism in M. Then we have inverse bijections

$$\{\langle s_1,..,s_l\rangle \in S_l\langle M\rangle ; s_l..s_1 = s\}$$

$$\downarrow \qquad \uparrow$$

$$\{([r_1],..,[r_{l-1}]) \in Qu_M(X)^{l-1} ; [id_X] \leq [r_1] \leq .. \leq [r_{l-1}] \leq [s]\} \quad ,$$

$$\langle s_1,..,s_l\rangle \mapsto ([s_1],[s_2 s_1],..,[s_{l-1}..s_1])$$

$$([r_1],..,[r_{l-1}]) \mapsto \langle r_1, r_2/r_1,..,r_{l-1}/r_{l-2}, s/r_{l-1}\rangle \quad .$$

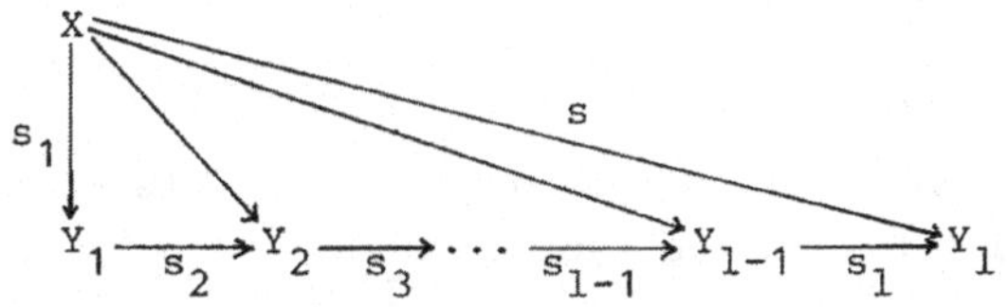

□

(1.39) COROLLARY: Let $s: X \to Y$ be an epimorphism in M. Then the dimension of s equals the maximal length of a chain in $Qu_M(X)$ from $[id_X]$ to $[s]$. □

We conclude that M satisfies the conditions (M1)-(M3).
Now let $\sim$ be an equivalence relation on M with the properties (~ 1)-(~ 3).

(1.40) COROLLARY: Condition (~ 4) holds if and only if, for all types t, t_1 and t_2, the number

$$\#\{[r] \in Qu_M(X) ; [id_X] \leq [r] \leq [s], \bar{r} = t_1, \overline{s/r} = t_2\}$$

is independent of the choice of the representative set. Then this number is the section coefficient $G(t;t_1,t_2)$. □

In what follows we also assume (~4). Hence $\underline{K}$,M,~ is a categorical structure.

Let X be an object in $\underline{K}$. Given an ascending sequence in $Qu_M(X)$

$$[r_0]\leq[r_1]\leq\ldots\leq[r_{l-1}]\leq[r_l] \quad , \quad l\geq 1 \quad ,$$

we call

$$(\overline{r_1/r_0},\overline{r_2/r_1},\ldots,\overline{r_l/r_{l-1}}) \in T^l$$

the <u>associated type sequence</u> . The ascending sequence $[r_0]\leq\ldots\leq[r_l]$ is a chain of length l if and only if all associated types are contained in T(1). By the following Lemma it suffices to consider ascending sequences with starting point $[id_X]$.

(1.41) LEMMA: Let $s_1:X\to Y_1$ be an epimorphism in M. Then the maps

$$\{[s_2]\in Qu_M(X);[s_1]\leq[s_2]\} \rightleftarrows Qu_M(Y_1)$$
$$[s_2] \longrightarrow [s_2/s_1]$$
$$[ss_1] \longleftarrow [s]$$

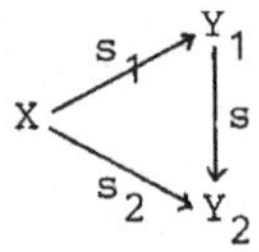

are mutually inverse order isomorphisms. Thus ascending sequences in $Qu_M(X)$ from $[s_1]$ to $[s_2]$ correspond to ascending sequences in $Qu_M(Y_1)$ from $[id_{Y_1}]$ to $[s_2/s_1]$. Under this correspondence, the associated type sequence is preserved. □

From Lemma (1.38) we derive a new interpretation of the higher section coefficients.

(1.42) PROPOSITION: Let $[s_1],[s_2]$ be quotient objects of X in M such that $[s_1]\leq[s_2]$, and let t be the type of s_2/s_1.
Then the number of ascending sequences in $Qu_M(X)$ from $[s_1]$ to $[s_2]$ with associated type sequence $(t_1,\ldots,t_l)\in T^l$, $l\geq 1$, is $G(t;t_1,\ldots,t_l)$. For every $l\geq 0$, there are exactly $G(t;l)$ chains of length l in $Qu_M(X)$ from $[s_1]$ to $[s_2]$. In particular, the maximal length of a chain from $[s_1]$ to $[s_2]$ equals dim(t). □

By Theorem (1.20), this implies the following result.

(1.43) COROLLARY: Suppose that the ground ring k has characteristic 0. Then the incidence algebra k(T;G) is graded by the sequence $(V(d))_{d\geq 0}$ if and only if the posets $Qu_M(X)$ satisfy the <u>(Jordan-Dedekind) chain condition</u>. □

Finally we establish the connection between the incidence algebra $k(T;G)$ and the incidence algebras of the posets $Qu_M(X)$, X an object in $\underline{K}$.

(1.44) THEOREM: For any commutative ring k, the map

$$1_X : k(T;G) \to I(Qu_M(X),k) \quad , \quad f \to 1_X(f) \quad ,$$

defined by

$$1_X(f)([s_1],[s_2]) = f(t) \quad \text{where } t \text{ is the type of } s_2/s_1 \quad ,$$

is a continuous k-algebra homomorphism.

In particular, let k be the ring of integers. Then 1_X preserves the zeta, Möbius and dimension function and, having extended the base ring to $k[D]$, also the characteristic function. For instance, if $\zeta,\mu,\dim,\chi$ denote the zeta, Möbius, dimension and characteristic function in $k(T;G)$ or $k[D](T;G)$ resp., we have

$$\zeta^2(t) = \#\{[r]\in Qu_M(X);[s_1]\leq[r]\leq[s_2]\} \quad ,$$

$$\mu(t) = \mu_{Qu_M(X)}([s_1],[s_2]) \quad \text{and}$$

$$(\zeta-\delta)^1(t) = \text{number of chains of length } 1 \text{ in } Qu_M(X) \text{ from } [s_1] \text{ to } [s_2] \quad .$$

Furthermore, if $Qu_M(X)$ satisfies the chain condition, then $\dim(t)$ is the <u>rank</u> of the interval $\{[r]\in Qu_M(X);[s_1]\leq[r]\leq[s_2]\}$, $\chi(t)$ is its <u>characteristic polynomial</u> ([Ro,64],p.343), and

$$D^{\dim}\zeta(t) = \Sigma_{i=0}^{\dim(t)} W(i)D^i$$

where the $W(i)$ are the <u>Whitney numbers</u> ([GK,78],p.24) which count the elements of rank i respectively. □

(end of the second case)

(1.45) REMARK: Let $\underline{K}$ be a skeletal-small category and let M be a class of monomorphisms in $\underline{K}$ with (S1)-(S3) (or a class of epimorphisms in $\underline{K}$ with (Q1)-(Q3)). We denote the subcategory of $\underline{K}$ consisting of the objects of $\underline{K}$ and the morphisms in M, by $\underline{M}$. Let <u>Lfp</u> be the category of locally finite posets together with proper maps (Given two posets P and Q, we call a map $\alpha:P\to Q$ proper if α is one-to-one, respects the order and preserves intervals). Then we have a functor

$$\underline{M} \to \underline{Lfp} \; , \quad \begin{array}{ccccc} Y_1 & \to & Sub_M(Y_1) & & [s_1] \\ {\scriptstyle s}\downarrow & & \downarrow{\scriptstyle Sub_M(s)} & & \downarrow \\ Y_2 & \to & Sub_M(Y_2) & & [ss_1] \end{array} \quad \left(\text{or} \quad \begin{array}{ccccc} X_1 & \to & Qu_M(X_1) & & [s_2s] \\ {\scriptstyle s}\downarrow & & \uparrow{\scriptstyle Qu_M(s)} & & \uparrow \\ X_2 & \to & Qu_M(X_2) & & [s_2] \end{array} \quad , \text{resp.}\right) .$$

In [DRS,72] it is shown that

$$\underline{\mathrm{Lfp}} \to \underline{\mathrm{Ass}}_k \ , \quad \begin{array}{ccc} P & \to & I(P,k) \\ \downarrow & & \uparrow {\scriptstyle I(\alpha,k)} \\ Q & \to & I(Q,k) \end{array} \quad \begin{array}{c} f \\ \uparrow \\ g \end{array} \ , \quad \begin{array}{c} f(p_1,p_2)=g(\alpha(p_1),\alpha(p_2)) \\ \\ , \end{array}$$

also is a functor. Here $\underline{\mathrm{Ass}}_k$ is the category of associative k-algebras with unit. Thus we end up with the functor

$$F\colon \underline{M} \to \underline{\mathrm{Ass}}_k \ , \ Y \to I(\mathrm{Sub}_M(Y),k) \quad (\text{or} \quad X \to I(\mathrm{Qu}_M(X),k) \quad \text{resp.}) \ .$$

Now let $\simeq$ be the isomorphism relation on M and consider the incidence algebra $k(I;\Gamma)$ of the categorical structure $\underline{K},M,\simeq$. Then it is not difficult to show that the algebra $k(I;\Gamma)$ together with the maps

$$1_Y\colon k(I;\Gamma) \to I(\mathrm{Sub}_M(Y),k) \quad (\text{or} \quad 1_X\colon k(I;\Gamma) \to I(\mathrm{Qu}_M(X),k) \quad \text{resp.})$$

is the inverse limit of the functor F.

(1.46) REMARK: In (1.26) and (1.27) we recapitulated the notion of the incidence algebra and the reduced incidence algebra of a locally finite poset. More generally, P.Doubilet, G.C.Rota and R.Stanley ([DRS,72]) have defined a "large incidence algebra" for a family of finite posets having some common features. In most examples where subobjects or quotient objects in a category are considered, their large incidence algebra coincides with the incidence algebra k(T;G) treated here.

§4. Functors and second cohomology

Let $\underline{K},M,\sim$ be a categorical structure. In this section we are concerned with the problem of determining when the incidence algebra k(T;G) is isomorphic to the large monoid algebra of a monoid with the finite factorization property, or to the incidence algebra of a locally finite poset.

Let $\underline{M}$ denote the subcategory of $\underline{K}$ consisting of the objects of $\underline{K}$ and the morphisms in M. We assume that the equivalence $\sim$ is given by

$$s_1 \sim s_2 \qquad \text{if and only if} \qquad Fs_1 = Fs_2$$

where $F\colon\underline{M} \to \underline{L}$ is a (covariant) functor and $\underline{L}$ is the category derived either from a monoid N with the finite factorization property as in (1.25), or from a locally finite poset P as in (1.26). In the first case, a functor F with ($\sim$1)-($\sim$4) is just a map $F\colon M \to N$ such that

(1.47) (i) $F(s)=1$ if and only if s is an isomorphism ,

(ii) $F(s_2s_1)=F(s_2)F(s_1)$ whenever s_1,s_2 are composable , and

(iii) for all y,z in the image of F, the number

$$\#\{<s_1,s_2>\in S_2<M>; s_2s_1=s, Fs_1=y, Fs_2=z\}$$

is independent of the choice of $s\in M$ with $F(s)=zy$.

Let Mor($\underline{L}$) be the set of morphisms in $\underline{L}$, and recall that S_1(Mor($\underline{L}$)) is the set of 1-simplices in $\underline{L}$. We identify the set of types T=M/~ with a subset of Mor($\underline{L}$) by $\bar{s} = Fs$. Since F is a functor, the section coefficients $G(t;t_1,t_2)$ vanish unless $t=t_2t_1$. Hence define

$$\theta: S_2(\text{Mor}(\underline{L})) \to \mathbb{N}_o$$

as follows: If r_1,r_2 and r_2r_1 are types, then $\theta(r_1,r_2)=G(r_2r_1;r_1,r_2)$. Otherwise, $\theta(r_1,r_2)=1$ if r_1 or r_2 is an identity, and $\theta(r_1,r_2)=0$ else.

(1.48) PROPOSITION: θ is a <u>normalized 2-cocycle</u> on $\underline{L}$. More explicitly,

(i) θ is normalized: If r_1 or r_2 is an identity, then $\theta(r_1,r_2)=1$.

(ii) for $(r_1,r_2,r_3)\in S_3(\text{Mor}(\underline{L}))$, the cocycle condition

$$\theta(r_1,r_2)\theta(r_2r_1,r_3) = \theta(r_1,r_3r_2)\theta(r_2,r_3) \quad \text{holds} .$$

Proof: This follows from the definition of θ and the corresponding properties of the section coefficients. □

(1.49) To proceed further, we need some background from homological algebra ([CE,56] or [HS,71]):

a. Let θ be an arbitrary normalized 2-cocycle on $\underline{L}$ with values in a commutative ring k. Then the topological k-module $k^{\text{Mor}(\underline{L})}$, endowed with the multiplication

$$fg(r) = \Sigma_{qp=r}\ \theta(p,q)f(p)g(q) \quad , \quad r\in\text{Mor}(\underline{L}) \quad ,$$

is an associative topological k-algebra with unit δ given by

$$\delta(r)=1 \text{ if r is an identity and } \delta(r)=0 \text{ otherwise} .$$

We denote this algebra by $k(\text{Mor}(\underline{L});\theta)$. For instance, if θ is the constant cocycle 1, then $k(\text{Mor}(\underline{L});1)$ is either the large monoid algebra of N or the incidence algebra of the poset P.

b. Let θ_1,θ_2 be two normalized 2-cocycles on $\underline{L}$ with values in k. Suppose that θ_1,θ_2 are <u>cohomologous</u>, i.e. it exists a (normalized) 1-chain $\varphi:\text{Mor}(\underline{L}) \to E(k)$ with values in the group of invertible elements of k such that

$$\varphi(r_2r_1)\theta_1(r_1,r_2) = \varphi(r_1)\varphi(r_2)\theta_2(r_1,r_2)$$

for all $(r_1,r_2)\in S_2(\text{Mor}(\underline{L}))$. Then the induced map

$$k(\text{Mor}(\underline{L});\theta_2) \to k(\text{Mor}(\underline{L});\theta_1) \ , \ f \to g \ , \ g(r)=f(r)/\varphi(r) \ ,$$

is an isomorphism of topological algebras. In particular,

$k(\mathrm{Mor}(\underline{L});\theta)$ is isomorphic to $k(\mathrm{Mor}(\underline{L});1)$ whenever θ is cohomologous to 1. Thus the monoid

$$H_n^2(\underline{L},k)$$

of the cohomology classes of normalized k-valued 2-cocycles on $\underline{L}$ becomes interesting. Here the product of two cocycles is defined by $\theta_1\theta_2(r_1,r_2) = \theta_1(r_1,r_2)\theta_2(r_1,r_2)$. In the sequel, we write $H_n^2(N,k)$ or $H_n^2(P,k)$ if $\underline{L}$ comes from the monoid N or from the poset P respectively.

c. Even $H_n^2(\mathbb{N}_o,\mathbb{Z})$ or $H_n^2(\mathbb{N}_o,\mathbb{Q})$ are difficult to characterize: If θ_1,θ_2 are cohomologous normalized 2-cocycles on $\mathbb{N}_o$, then their vanishing sets V_1,V_2 in $\mathbb{N}_o\times\mathbb{N}_o$ are equal. So one might try first to classify all possible vanishing sets. But it turns out that there are plenty of subsets V of $\mathbb{N}_o\times\mathbb{N}_o$ such that the function

$$\chi: \mathbb{N}_o\times\mathbb{N}_o \to \mathbb{Z}\ ,\ \chi=0 \text{ on } V \text{ and } \chi=1 \text{ outside } V\ ,$$

is a normalized cocycle on $\mathbb{N}_o$. In the following examples, W is the complement of V in $\mathbb{N}\times\mathbb{N}$.

a single point: $W_1=\{(m,n)\}$, $m,n\in\mathbb{N}$

a triangle: $W_2=\{(m,n)\in\mathbb{N}\times\mathbb{N};m+n\leq r\}$, $r\in\mathbb{N}$

a lattice: $W_3=S\times S$, S a subset of $\mathbb{N}$ closed under addition

a diagonal sequence: $W_4=\{(2^{i+1}-1,2^{i+1}-1);i=0,1,..\}$

twins of prime numbers:

$$W_5=\{(m,2);m\geq 5 \text{ and } m,m+2 \text{ are prime numbers}\}\cup$$
$$\cup\{(2,n);n\geq 5 \text{ and } n,n+2 \text{ are prime numbers}\}\ .$$

Moreover, every normalized 2-cocycle γ on the cyclic group $\mathbb{Z}/\mathbb{Z}l$ induces a normalized 2-cocycle θ on $\mathbb{N}_o$ by $\theta(m,n)=\gamma(\bar{m},\bar{n})$ where $\bar{m},\bar{n}$ are the residue classes of m,n modulo l. Here the vanishing set of θ is l-periodic.

For a detailed study of normalized 2-cocycles on $\mathbb{N}_o$ with zeros, we refer the reader to [Dür,81].

d. Matters become easier if the normalized 2-cocycle θ on $\underline{L}$ only takes values in the group E(k) of invertible elements of k since the second cohomology group

$$H^2(\underline{L},A)\ ,\ A \text{ an abelian group}\ ,$$

is known in many cases. Notice that every 2-cocycle on $\underline{L}$ with values in A is cohomologous to a normalized cocycle.

e. If $N=\mathrm{Wo}(I)$ is the free monoid of words over a set I of letters, then $H^2(\mathrm{Wo}(I),A)=0$ by [CE,56],pp.192, and $k(\mathrm{Wo}(I);\theta)$ is isomorphic to the topological word algebra over the alphabet I $k\langle\langle I\rangle\rangle = k(\mathrm{Wo}(I);1)$.

f. If $N=\mathbb{N}_o$ is the additive monoid of non-negative integers, again $H^2(\mathbb{N}_o,A)=0$ by [CE,56],pp.192, and $k(\mathbb{N}_o;\theta)$ is isomorphic to the power series algebra in one variable z $k[[z]] = k(\mathbb{N}_o;1)$. Because of its importance in combinatorics, we discuss this case in detail: $H^2(\mathbb{N}_o,E(k))=1$ implies that there is a function $\varphi: \mathbb{N}_o \to U(k)$ such that

$$\theta(m,n) = \frac{\varphi(m+n)}{\varphi(m)\varphi(n)} \quad \text{for all } m,n\in \mathbb{N}_o .$$

A straight-forward calculation shows that one such φ is given by

$$\varphi(0)=\varphi(1)=1 \quad \text{and} \quad \varphi(n) = \Pi_{i=1}^{n-1} \theta(i,1) \quad \text{for } n\geq 2 .$$

Then we have an algebra isomorphism

$$k(\mathbb{N}_o;\theta) \to k[[z]] , \ f \to \Sigma_{n=0}^{\infty} f(n)z^n/\varphi(n) .$$

As $E(\mathbb{Z})=\{1,-1\}$ is too small, usually $k=\mathbb{Q}$, $\mathbb{R}$ or $\mathbb{C}$. In [Co,74] and in [St,78], various kinds of generating functions

$$\Sigma_{n=0}^{\infty} f(n)z^n/\varphi(n)$$

are considered:

ordinary g.f. : $\varphi(n)=1$

exponential g.f. : $\varphi(n)=n!$

Eulerian g.f. : $\varphi(n)=[n]!=[n][n-1]..[1]$, $[n]=(q^n-1)/(q-1)$

doubly-exponential g.f. : $\varphi(n)=(n!)^2$

chromatic g.f. : $\varphi(n)=q^{n(n-1)/2}n!$

For each kind, the cocycle θ, $\theta(m,n)=\varphi(m+n)/\varphi(m)\varphi(n)$ for $m,n\in \mathbb{N}_o$, has integer values and also admits a combinatorial interpretation (see [St,78]). We will come back to this observation in the next item.

g. More generally, let $N=\mathbb{N}_o(I)$ be the free commutative monoid generated by a set I. The Koszul complex calculations of [CE,56], pp.192, show that every 2-cocycle γ on $\mathbb{N}_o(I)$ with values in A is cohomologous to a bilinear alternating map $b: \mathbb{N}_o(I)\times \mathbb{N}_o(I) \to A$. Hence $k(\mathbb{N}_o(I);\theta)$ is isomorphic to a "skew power series algebra" (in non-commuting indeterminates z_i, $i\in I$).
But if θ is symmetric and thus cohomologous to 1, then $k(\mathbb{N}_o(I);\theta)$ is isomorphic to the power series algebra $k(\mathbb{N}_o(I);1) = k[[z_i;i\in I]]$ in the variables z_i, $i\in I$, by

$$k(\mathbb{N}_o(I);\theta) \to k[[z_i;i\in I]] , \ f \to \Sigma_{n\in \mathbb{N}_o(I)} f(n)z^n/\varphi(n)$$

$(z^n = \Pi_{i\in I} z_i^{n(i)})$.

Since only 1 and -1 are invertible in $\mathbb{Z}$, this will be applied when $k=\mathbb{Q}$, $\mathbb{R}$ or $\mathbb{C}$.

Generating functions of the form $\Sigma_n f(n)z^n/\varphi(n)$ arising in enumeration problems seem to have the following property:

The cocycle θ, $\theta(m,n)=\varphi(m+n)/\varphi(m)\varphi(n)$ for $m,n\in \mathbb{N}_o(I)$, has all its values in the set of positive integers.

Then the ring $\mathbb{Z}(\mathbb{N}_o(I);\theta)$ is isomorphic to the subring

$$\{ \Sigma_n \lambda_n z^n/\varphi(n) \ ; \ \lambda_n\in\mathbb{Z} \text{ for all } n \}$$

of the power series algebra $\mathbb{Q}[[z_i;i\in I]]$. Furthermore, different cocycles with values in $\mathbb{N}$ determine different rings of power series, e.g. the rings of ordinary, exponential or Eulerian generating functions (with integer coefficients). Examples for the multivariate case appear throughout this book. Observe that this distinction is impossible when $k=\mathbb{Q}$, $\mathbb{R}$ or $\mathbb{C}$.

In particular, if I is the set of prime numbers, then $\mathbb{N}_o(I)$ is isomorphic to the monoid $\mathbb{N}$ of positive integers with multiplication by the map $n \to \Pi_{p\in I} p^{n(p)}$, and the preceding considerations apply. Here the power series algebra $\mathbb{C}[[z_p;p\in I]]$ can be viewed as the algebra of formal Dirichlet series with complex coefficients $\Sigma_{n=1}^{\infty} \lambda_n n^{-s}$ via the transformation $z_p \to p^{-s}$, $p\in I$.

h. If P is a countable directed set, then $H^2(P,A)=0$ by [Roos,61], and $k(\{(x,y)\in P\times P;x\leq y\};\theta)$ is isomorphic to the algebra of upper triangular $P\times P$-matrices over k.

We now return to the situation studied in the beginning. For any ground ring k, the cocycle θ obtained from the section coefficients $G(t;t_1,t_2)$ defines a normalized 2-cocycle on $\underline{L}$ with values in k which we also denote by θ.

(1.50) PROPOSITION: The incidence algebra $k(T;G)$ is isomorphic to the factor algebra $k(\mathrm{Mor}(\underline{L});\theta)/J$ where J is the closed ideal of functions on $\mathrm{Mor}(\underline{L})$ vanishing on T. If the functor F is surjective on morphisms, then $k(T;G) = k(T;\theta)$.

Proof: By the definition of θ, $k(\mathrm{Mor}(\underline{L});\theta) \to k(T;G)$, $g \to g|T$, is a surjective homomorphism of topological k-algebras. □

(1.51) COROLLARY: If the functor F is surjective on morphisms, if all numbers

$$\theta(t_1,t_2) = \#\{\langle s_1,s_2\rangle\in S_2\langle M\rangle;s_2s_1=s,Fs_1=t_1,Fs_2=t_2\}$$

where $s\in M$ with $Fs=t_2t_1$, are non-zero and if the second cohomology group $H^2(\underline{L},E(\mathbb{Q}))$ is trivial, then the incidence algebra $\mathbb{Q}(T;G)$ is isomorphic either to a large monoid algebra or to the incidence algebra of a poset, depending on $\underline{L}$. □

(1.52) REMARK: In the sections "7.Algebras of Dirichlet type, 8.Algebras of full binomial type and 9.Algebras of triangular type" of the article [DRS,72], the authors study three kinds of locally finite posets and their reduced incidence algebras where the equivalence ~ is defined by a functor $F:\underline{M} \to \underline{L}$. In section 7 and 8 the category $\underline{L}$ is derived from the monoid $\mathbb{N}_0$ or $\mathbb{N}$ resp., whereas in section 9 $\underline{L}$ comes from the posets $\{0,1,..,N\}$ or $\mathbb{N}_0$ in their natural order. In the three sections cohomological arguments appear often (compare with (1.49), g, f and h), but they are not mentioned explicitly. For combinatorial applications, see also [St,76] and [St,78].

§5. Examples

In this section we consider, for various combinatorial categories, the incidence algebra associated with the posets of subobjects by the formalism of §3. The examples given here are collected from the literature where they have been studied per se. New examples will be presented in chapter IV.

(1.53) Subsets of finite sets and exponential power series:

Let $\underline{K}$ be the category $\underline{Sf}$ of finite sets. In $\underline{Sf}$ the monomorphisms are the injective maps. Let M be the class of all monomorphisms in $\underline{Sf}$. A subobject $[s:X\to Y]$ of a finite set Y usually is identified with the subset $s(X)$ of Y. Thus the poset Sub(Y) is the power set of Y ordered by inclusion. Define an equivalence relation ~ on M by

$$s_1 \sim s_2 \quad \text{if and only if} \quad F(s_1) = F(s_2)$$

where the map $F:M \to \mathbb{N}_0$ is given by $F(s) = \#Y - \#X$ for $s:X\to Y$. Then F satisfies all conditions in (1.47). As in §4 we identify $T = \mathbb{N}_0$. By Corollary (1.33) the section coefficients

$$G(m+n;m,n) = \theta(m,n) = \binom{m+n}{m} = \frac{(m+n)!}{m!n!}$$

are the binomial coefficients, and Corollary (1.51) yields the isomorphism of topological $\mathbb{Q}$-algebras

$$\rho: \mathbb{Q}(T;G) \to \mathbb{Q}[[z]] \ , \ f \to \Sigma_{n=0}^{\infty} f(n)z^n/n! \ .$$

As $\rho(\zeta)=\exp(z)$ and hence $\rho(\mu)=\exp(-z)$, we find that $\mu(n)=(-1)^n$, $n\geq 0$. In order to apply Theorem (1.37), let X,Z be two subsets of the finite set Y such that $X\subset Z$. If $s_1:X\to Y$, $s_2:Z\to Y$ are the inclusion maps, then also $s_2\backslash s_1:X\to Z$ is the inclusion map and has type $n = \#Z - \#X$.

By (1.37) we obtain the basic result

$$\mu_{Sub(Y)}(X,Z) = \mu(n) = (-1)^n \qquad ([Ro,64]) .$$

Similarly, using the isomorphism $\rho\hat{\otimes}_{\mathbb{Q}}A\colon A(T;G) \to A[[z]]$ where the base ring has been extended to $A=\mathbb{Q}[D]$, the characteristic polynomial and the Whitney numbers of the Boolean algebra of subsets of Y can be computed by algebraic means.

(1.54) Subspaces of finite vector spaces and Eulerian power series:

Let $\mathbb{F}$ be a finite field with q elements and let $\underline{K}$ be the category $\underline{Vf}_{\mathbb{F}}$ of finite-dimensional vector spaces over $\mathbb{F}$. In $\underline{Vf}_{\mathbb{F}}$ the monomorphisms are the injective $\mathbb{F}$-linear maps. Let M be the class of all monomorphisms in $\underline{Vf}_{\mathbb{F}}$. Since a subobject $[s:X\to Y]$ of a finite $\mathbb{F}$-vector space Y can be identified with the subspace s(X) of Y, the poset Sub(Y) is the projective geometry of Y. Define an equivalence relation $\sim$ on M by

$$s_1 \sim s_2 \qquad \text{if and only if} \qquad F(s_1) = F(s_2)$$

where $F:M \to \mathbb{N}_0$, $F(s)= \dim_{\mathbb{F}}Y - \dim_{\mathbb{F}}X$ for $s:X\to Y$, and $\dim_{\mathbb{F}}$ denotes the dimension over $\mathbb{F}$. Again we are in the situation of §4. Here $T= \mathbb{N}_0$, the section coefficients

$$G(m+n;m,n) = \theta(m,n) = \begin{bmatrix} m+n \\ m \end{bmatrix} = \frac{[m+n]!}{[m]![n]!}$$

$$(\text{ where } [l]!=[l][l-1]..[1] , [l]=(q^l-1)/(q-1))$$

are the Gaussian or q-binomial coefficients, and we have an isomorphism

$$\rho\colon \mathbb{Q}(T;G) \to \mathbb{Q}[[z]] , \; f \mapsto \Sigma_{n=0}^{\infty} f(n)z^n/[n]! \quad .$$

This is the "projective form" of [GR,70],p.246, which converges to the corresponding isomorphism for sets as $q\to 1$. $\rho(\zeta)= \Sigma_{n=0}^{\infty} z^n/[n]!$ is the q-exponential function of J.Cigler ([Ci,82]). From $\rho(\mu)= 1/e(z)= \Sigma_{n=0}^{\infty}(-1)^n q^{n(n-1)/2}z^n/[n]!$, a basic relation in the theory of q-identities, we read off that $\mu(n)=(-1)^n q^{n(n-1)/2}$, $n\geq 0$. If $X\subset Z$ are two subspaces of the finite $\mathbb{F}$-vector space Y, then Theorem (1.37) implies the following result, due to P.Hall:

$$\mu_{Sub(Y)}(X,Z) = \mu(n) = (-1)^n q^{n(n-1)/2} , \quad n= \dim_{\mathbb{F}}Z - \dim_{\mathbb{F}}X .$$

As a further application of the isomorphism ρ, we set up the Eulerian generating function of the Galois numbers G_n which count the subspaces of $\mathbb{F}^n$ respectively. Since $\#Sub(\mathbb{F}^n)=\zeta^2(n)$ by (1.37) and $\rho(\zeta^2)=\rho(\zeta)^2$, it follows that

$$\Sigma_{n=0}^{\infty} G_n \, z^n/[n]! = e(z)^2 \quad .$$

(1.55) Subwords and the shuffle algebra:

Let A be an arbitrary set. By

$$Wo(A) = \{w:[n]\to A; n=0,1,..\}$$

we denote the set of words over the alphabet A. Here $[n]=\{1,..,n\}$, in particular $[0]=\emptyset$, the empty set. Usually a word $w:[n]\to A$ is written as $w(1)..w(n)$.

We consider the following category K: The objects in K are the words over A. Given two words $v:[m]\to A$ and $w:[n]\to A$, a morphism $s:v\to w$ is a triple (v,β,w) where $\beta:[m]\to[n]$ is a monotonous map such that $w\cdot\beta=v$. The composition of two morphisms (u,α,v) and (v,β,w) is defined by $(v,\beta,w)(u,\alpha,v)=(u,\beta\alpha,w)$.

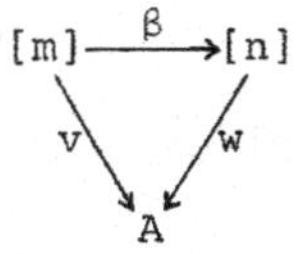

Obviously, there are no isomorphisms in K except the identities, and a morphism $s=(v,\beta,w)$ is a monomorphism if and only if β is one-to-one. Thus a subobject of a word $w:[n]\to A$ can be identified with the unique representative $s=(v,\beta,w)$. For instance, if $A=\{a,b\}$, then $r=(ab,\alpha,abba), \alpha(1)=1, \alpha(2)=2$, and $s=(ab,\beta,abba), \beta(1)=1, \beta(2)=3$, are two different subobjects of abba. Following [Lo,82], we call subobjects subsequences and their domains subwords. So r,s from above determine the same subword ab of abba. Given a monomorphism $s:v\to w$, a new word $w-s(v)$ is obtained from the word w by deleting the subsequence s.

Now let M be the class of all monomorphisms in K and define the equivalence relation $\sim$ by

$$(s_1:v_1\to w_1)\sim(s_2:v_2\to w_2) \quad \text{if and only if} \quad w_1-s_1(v_1)=w_2-s_2(v_2) \quad .$$

Then K,M,$\sim$ are a categorical structure. For (S3) observe that every set Sub(w) is finite, and ($\sim$4) is easily checked by Corollary (1.33).

We identify the set of types $T=M/\sim$ with Wo(A) by $\overline{s:v\to w} = w-s(v)$.

If w,u,v are words, we see from (1.33), choosing s the empty subsequence of w, that the section coefficient G(w;u,v) counts all subsequences $r:u\to w$ of w such that $w-r(u)=v$. For example, G(abba;ab,ba)=2. In particular, the "generalized binomial coefficient"

$$\binom{w}{u} = \Sigma_v\, G(w;u,v)$$

tells how often u is a subword of w (When A contains only one letter, this is the usual binomial coefficient).

Hence the incidence algebra $\mathbb{Z}$(Wo(A);G) is the topological shuffle algebra over the alphabet A, i.e. the $\mathbb{Z}$-module $\mathbb{Z}\langle\langle A\rangle\rangle$ of all formal series over A with integer coefficients, endowed with the shuffle product and the topology of coefficientwise convergence ([Lo,82]).

Since the section coefficients are symmetric, the incidence algebra

$\mathbb{Z}(Wo(A);G)$ is commutative.

For every word w, the poset of subobjects of w clearly is isomorphic to the power set of $[l(w)]$ where $l(w)$ denotes the length of w. From (1.53) and Theorem (1.37) it follows that the Möbius function μ in $\mathbb{Z}(Wo(A);G)$ is given by

$$\mu(w) = (-1)^{l(w)} \quad , \quad w \in Wo(A) \ .$$

Thus Möbius inversion in $\mathbb{Z}(Wo(A);G)$ reads: For $f,g \in \mathbb{Z}(Wo(A);G)$,

$$g(w) = \Sigma_v \binom{w}{v} f(v) \qquad \text{for all } w \in Wo(A)$$

if and only if

$$f(w) = \Sigma_v (-1)^{l(w)-l(v)} \binom{w}{v} g(v) \qquad \text{for all } w \in Wo(A) \ .$$

For instance, let $f=e(u)$ for any $u \in Wo(A)$, and define g by the first equation. Then the second equation yields the orthogonal relation

$$\Sigma_v (-1)^{l(w)+l(v)} \binom{w}{v}\binom{v}{u} = \delta_{u,v} \qquad \text{for all } w \in Wo(A)$$

of [Lo,82],p.124.

(1.56) <u>Subgroups of finite abelian groups and an algebra of S.Delsarte:</u>

Let <u>K</u> be the category <u>Abf</u> of finite abelian groups, and let M be the whole class of monomorphisms. In <u>Abf</u> the monomorphisms are the injective group homomorphisms, and a subobject $[s:X \to Y]$ of a finite abelian group Y is currently identified with the subgroup $s(X)$ of Y. Thus the poset $Sub(Y)$ is the lattice of subgroups of Y. Define the equivalence relation $\sim$ on M by $(s_1:X_1 \to Y_1) \sim (s_2:X_2 \to Y_2)$ if and only if the factor groups $Y_1/s_1(X_1)$ and $Y_2/s_2(X_2)$ are isomorphic.

Then it is easy to check that all conditions are satisfied. Obviously, one can identify the types $t \in T=M/\sim$ with the isomorphism classes of finite abelian groups by $\overline{s:X \to Y}$ = isomorphism class of $Y/s(X)$.

In particular, a representative of type t is the homomorphism $s:0 \to Y$ where the group Y belongs to the isomorphism class t. From Corollary (1.33) we see that, for types t,t_1,t_2 and a finite abelian group Y of type t, the section coefficient $G(t;t_1,t_2)$ counts the subgroups X of Y such that X has type t_1 and Y/X has type t_2.

The incidence algebra $\mathbb{C}(T;G)$ has been studied by S.Delsarte ([Del,48]). Since the section coefficients are symmetric, the algebra $\mathbb{C}(T;G)$ is commutative. In his article, Delsarte also determines the Möbius function μ in $\mathbb{Z}(T;G)$:

$$\mu(t) = \Pi_{p \text{ prime}} (-1)^{n(p)} p^{n(p)(n(p)-1)/2}$$

if t is the type of a product $\Pi_{p \text{ prime}} (\mathbb{Z}_p)^{n(p)}$ where $\mathbb{Z}_p$ is the cyclic group of order p and almost all n(p) are zero, and

$$\mu(t) = 0 \qquad \text{otherwise} .$$

Finally, we mention two special cases of (1.56) where full subcategories of Abf are considered.

(1.57) Subgroups of abelian p-groups and the algebra of P.Hall:

As the subgroup lattice of a finite abelian group is the product of the subgroup lattices of its p-Sylow subgroups, it is reasonable to consider the following categorical structure: Given a prime p, let K be the category of abelian p-groups, let M be the class of monomorphisms in K and let the equivalence relation ~ on M be defined as in (1.56). But any abelian p-group Y is a direct sum of cyclic subgroups, of orders $p^{\lambda_1}, p^{\lambda_2}, \ldots, p^{\lambda_r}$ say, where we may suppose that $\lambda_1 \geq \lambda_2 \geq \ldots \geq \lambda_r$, and the sequence of exponents $\lambda = (\lambda_1, \ldots, \lambda_r)$ is a partition of a natural number which describes Y up to isomorphism. We hence set

$$T = \{\lambda;\ \lambda \text{ a partition of a natural number}\} \quad .$$

Then the section coefficients $G(\lambda;\alpha,\beta)$ counting the subgroups X of an abelian p-group Y of type λ such that X has type α and Y/X has type β, are the Hall polynomials $g^{\lambda}_{\alpha\beta}(p)$ defined in [Ha,59]. Hall's algebra A(p) is isomorphic to the subalgebra of $\mathbb{C}(T;G)$ consisting of all functions which are non-zero on only finitely many types. A detailed exposition of Hall's theory stressing the connection with symmetric functions is given in the book of I.G.Macdonald ([Ma,79]). Therein, the Hall algebra is mapped isomorphically onto the algebra of symmetric functions in infinitely many variables, thus relating the subgroup lattices of abelian p-groups to the multiplication rule of the Hall-Littlewood functions. Consequently, we can realize the incidence algebra $\mathbb{Q}(T;G)$ as an algebra of formal power series by

$$\rho: \mathbb{Q}(T;G) \to \mathbb{Q}[[z_1,z_2,\ldots]] \ , \ f \to \Sigma_\lambda \ f(\lambda)P_\lambda(z;p^{-1})/p^{n(\lambda)} \quad .$$

Here $P_\lambda(z;t)$ is the Hall-Littlewood function associated with the partition $\lambda=(\lambda_1,\lambda_2,\ldots)$, and $n(\lambda)=\Sigma_{i\geq 1}(i-1)\lambda_i$. For example, $\rho(\zeta) = \Sigma_\lambda P_\lambda(z;p^{-1})/p^{n(\lambda)} = \Pi_{i\geq 1}(1-z_i)^{-1}$, $\rho(\zeta^2) = \Pi_{i\geq 1}(1-z_i)^{-2}$, and

$$\Pi_{i\geq 1} \ (1-z_i)^{-2} = \Sigma_\lambda \ \zeta^2(\lambda) \ P_\lambda(z;p^{-1})/p^{n(\lambda)}$$

where $\zeta^2(\lambda)$ counts the subgroups of an abelian p-group of type λ. More generally, in place of abelian p-groups Macdonald considers modules of finite length over a discrete valuation ring with finite residue field.

(1.58) Subgroups of finite cyclic groups and formal Dirichlet series:

Here K is the category of finite cyclic groups, M is the class of all monomorphisms in K, and the equivalence ~ is defined as in (1.56). But

$$s_1 \sim s_2 \quad \text{if and only if} \quad F(s_1) = F(s_2)$$

where $F:M \to \mathbb{N}$, $F(s) = \#Y/\#X$ for $s:X \to Y$, satisfies the conditions (1.47). The types can be identified with positive integers, the section coefficients are

$$G(mn;m,n) = \theta(m,n) = 1 \quad , \quad m,n \in \mathbb{N} \quad ,$$

and the incidence algebra $\mathbb{C}(T;G)$ is isomorphic to the algebra of formal Dirichlet series with complex coefficients by the map

$$f \to \Sigma_{n=1}^{\infty} f(n)n^{-s} \quad .$$

Thus $\zeta \in \mathbb{Z}(T;G)$ corresponds to Riemann's zeta function $\Sigma_{n=1}^{\infty} n^{-s}$ whereas $\mu \in \mathbb{Z}(T;G)$ is the classical Möbius function from number theory.

CHAPTER II

PARTITIONS, FUNCTORS AND EXPONENTIAL FORMULAS

§1. Partitions in categories

Let $\underline{K}$ be a skeletal-small category with the properties ($\underline{K}$1)-($\underline{K}$5) stated below:

($\underline{K}$1) In $\underline{K}$ exists an object 0 such that, for any object X, there is a unique morphism $0 \to X$.

Such objects are called initial. Clearly all initial objects in $\underline{K}$ are isomorphic. Given an object Y in $\underline{K}$, we call a subobject $[r:Z \to Y]$ initial if Z is initial. Then it is the least subobject of Y.

($\underline{K}$2) For any two objects X_1, X_2 in $\underline{K}$, there exists an object X and monomorphisms $u_1:X_1 \to X$, $u_2:X_2 \to X$ satisfying the following condition: Given two morphisms $s_1:X_1 \to Y$, $s_2:X_2 \to Y$ in $\underline{K}$, there is a unique morphism s such that $s_1 = su_1$ and $s_2 = su_2$.

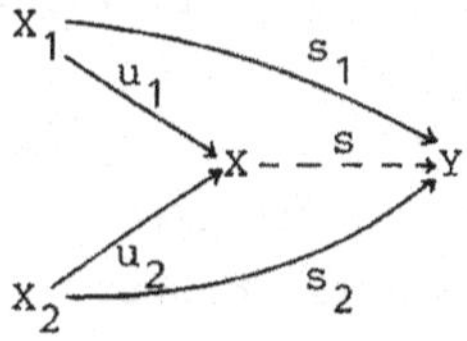

We indicate this property of X and u_1, u_2 by writing $X = (X_1, u_1) \oplus (X_2, u_2)$. If also $X' = (X_1, u_1') \oplus (X_2, u_2')$, then there is a unique isomorphism $b:X \to X'$ such that $u_i' = bu_i$ for $i=1,2$. Among all X and $u_1:X_1 \to X$, $u_2:X_2 \to X$ such that $X = (X_1, u_1) \oplus (X_2, u_2)$, we choose once for all a copy $X_1 \oplus X_2$ which is called the direct sum of X_1 and X_2 with the canonical injections $X_1 \to X_1 \oplus X_2$ and $X_2 \to X_1 \oplus X_2$. They are, in particular, monomorphisms.

($\underline{K}$3) The direct sum is disjoint: If X is the direct sum of X_1, X_2 with the canonical injections $u_1:X_1 \to X$, $u_2:X_2 \to X$ and if [r] is a subobject of X contained both in $[u_1]$ and $[u_2]$, then [r] is the initial subobject of Y.

From the conditions ($\underline{K}$1)-($\underline{K}$3) it follows inductively that in $\underline{K}$ finite direct sums exist and are unique up to isomorphism, that the canonical injections are monomorphisms and that finite direct sums are disjoint

([BD,68] or [ML,71]). As the empty direct sum the initial object 0 is chosen.

Let X be an object in $\underline{K}$. By a decomposition of X we understand an isomorphism $b:X \to \oplus_{i\in I}X_i$ where I is a finite index set. The decomposition is proper if I has at least two elements and no X_i is initial. X is called indecomposable if X is not initial and admits no proper decompositions.

(2.1) DEFINITION: Let Y be an object in $\underline{K}$.

(i) A partition of Y in $\underline{K}$ is a finite set

$$\pi = \{[r_i:X_i \to Y]; i=1,..,l\} \qquad (l \geq 0)$$

of non-initial subobjects of Y, called the blocks of π, such that

$$Y = \oplus_i (X_i, r_i) .$$

If each X_i is indecomposable, we call π a Krull-Schmidt partition and $Y \simeq \oplus_i X_i$ a Krull-Schmidt decomposition of Y (In the sequel, we will abbreviate "Krull-Schmidt" by KS).

(ii) Let Pa(Y) denote the set of partitions of Y. Then the automorphism group of Y acts on Pa(Y) by

$$\mathrm{Aut}(Y) \times \mathrm{Pa}(Y) \to \mathrm{Pa}(Y) , \quad (b,\pi) \to b\pi ,$$

where $b\pi = \{[br_i:X_i \to Y]; i=1,..,l\}$ for $\pi = \{[r_i:X_i \to Y]; i=1,..,l\}$.

The most important condition for $\underline{K}$ is

($\underline{K}$4) Every object in $\underline{K}$ has a Krull-Schmidt partition which is unique up to the action of the automorphism group.

(2.2) DEFINITION: We call a skeletal-small category with the properties ($\underline{K}$1)-($\underline{K}$4) a Krull-Schmidt category. This name stems from the theorem of Krull-Schmidt ([At,56]) which states sufficient conditions for ($\underline{K}$4) in an abelian category. Obviously, finite products of KS-categories are again KS-categories.

Finally, we assume

($\underline{K}$5) For every object X in $\underline{K}$, the automorphism group Aut(X) is finite.

Now fix a system $\underset{\sim}{P}$ of representatives of the isomorphism classes of indecomposable objects in $\underline{K}$, and let

$$\mathbb{N}_o(\underset{\sim}{P}) = \{n \in \mathbb{N}_o^{\underset{\sim}{P}}; \ n(P) \neq 0 \text{ for only finitely many } P \in \underset{\sim}{P}\}$$

be the free commutative monoid with basis $(\varepsilon(P); P \in \underset{\sim}{P})$.

(2.3) DEFINITION: Let X be an object in $\underline{K}$. By ($\underline{K}$4) there is a unique $n \in \mathbb{N}_o(\underset{\sim}{P})$ such that

$$X \simeq \oplus_{P \in \underset{\sim}{P}} \underbrace{(P \oplus \ldots \oplus P)}_{n(P)\ \text{times}} = \oplus_P n(P) P \quad .$$

We call $n = n_X$ the Krull-Schmidt type of X. When $\underset{\sim}{P}$ has only one member, we identify $\mathbb{N}_o(\underset{\sim}{P})$ with $\mathbb{N}_o$ and view KS-types as natural numbers. Evidently, $n_X = 0$ if and only if X is initial, and $n_{X \oplus Y} = n_X + n_Y$ for two objects X,Y in $\underline{K}$. So $\mathbb{N}_o(\underset{\sim}{P})$ is isomorphic to the monoid of isomorphism classes of objects in $\underline{K}$ where addition is induced from direct sum. Furthermore, by ($\underline{K}$5) we can define

$$a: \mathbb{N}_o(\underset{\sim}{P}) \to \mathbb{N} \ , \ a(n) = \text{order of } \operatorname{Aut}(\oplus_P n(P) P) \ .$$

Our next aim is to count partitions in $\underline{K}$. Before, we look at some Krull-Schmidt categories in combinatorics.

(2.4) EXAMPLES:

a. In the category $\underline{Sf}$ of finite sets, the only initial object is the empty set $\emptyset$ and direct sum is disjoint union. Hence the indecomposable objects in $\underline{Sf}$ are the sets with a single element. A subobject $[s:X \to Y]$ of a finite set Y is identified with the subset s(X) of Y. Then definition (2.1) coincides with the usual definition of a partition of a finite set, and the (even unique) KS-partition of an object X is

$$\pi = \{\{x\}; x \in X\} \ .$$

Since all indecomposable objects are isomorphic, the system of representatives $\underset{\sim}{P}$ has only one member P which is a set with a single element. Thus the KS-type of a finite set X equals the number of elements of X because $X \simeq P \dot{\cup} .. \dot{\cup} P$ (n times, $n = \# X$), and $a(n) = n!$ is the factorial.

b. Given a positive integer r, let $\underline{Sf}^{(r)}$ be the full subcategory of the product category $\underline{Sf}^r$ whose objects $X = (X_1, .., X_r)$ satisfy $\# X_1 = .. = \# X_r$. Here the only initial object is $(\emptyset, .., \emptyset)$, and direct sum is disjoint union in each component. An object $X = (X_1, .., X_r)$ is indecomposable if and only if $\# X_1 = .. = \# X_r = 1$. We identify a subobject $[s:X \to Y]$ of $Y = (Y_1, .., Y_r)$ with $(s_1(X_1), .., s_r(X_r))$ where each $s_i(X_i)$ is a subset of Y_i. Then the partitions of $Y = (S, .., S)$ in $\underline{Sf}^{(r)}$, S a finite set, are just the "r-partitions" of S introduced by R.P.Stanley in [St,78]. Again, there is exactly one indecomposable object in $\underline{Sf}^{(r)}$ up to isomorphism. Hence the KS-type of $X = (X_1, .., X_r)$ is the common cardinality of $X_1, .., X_r$, and $a(n) = (n!)^r$.

c. Let $\mathbb{F}$ be a finite field with q elements. In the category $\underline{Vf}_{\mathbb{F}}$ of finite-dimensional vector spaces over $\mathbb{F}$, the initial objects are the vector spaces containing only the zero-vector. Since direct sum is the usual one, an object Z is indecomposable in $\underline{Vf}_{\mathbb{F}}$ if and only if Z has dimension 1 over $\mathbb{F}$. A subobject $[s:X\to Y]$ of a finite $\mathbb{F}$-vector space Y is identified with the subspace s(X) of Y. Then a partition of Y is a set of non-zero subspaces of Y such that Y is their direct sum. In $\underline{Vf}_{\mathbb{F}}$ however, KS-partitions of objects are not unique as in the previous examples, but they are unique up to the action of the general linear group. Choosing $\underset{\sim}{P}=\{F\}$, we see from $X \simeq \mathbb{F}\oplus..\oplus\mathbb{F}$ (n times, $n=\dim_{\mathbb{F}}X$) that the KS-type of X equals the dimension of X over $\mathbb{F}$. Hence $a(n)=(q^n-1)(q^n-q)..(q^n-q^{n-1})$ is the order of the general linear group in $\mathbb{F}^n$.

d. By a (finite undirected simple) graph we understand a pair (V,E) where V (the vertex set) is a finite set and E (the edge set) is a set of two-element subsets of V. A graph homomorphism $s:(V,E)\to(W,F)$ is a map $s:V\to W$ between the vertex sets preserving edges, i.e. if $\{v_1,v_2\}\in E$, then $\{s(v_1),s(v_2)\}\in F$. In the category $\underline{Gf}$ of graphs, the only initial object is the empty graph $(\emptyset,\emptyset)$ and direct sum is disjoint union of the vertex sets and edge sets. The indecomposable objects in $\underline{Gf}$ are the non-empty connected graphs, and a subobject $[s:(V,E)\to(W,F)]$ of a graph (W,F) can be identified with the subgraph (s(V),s(E)) of (W,F) where $s(E)=\{\{s(e_1),s(e_2)\};\{e_1,e_2\}\in E\}$. Hence the unique KS-partition of a graph is the set of its connected components. In the combinatorial literature, (finite undirected simple) graphs often are called "labeled" graphs whereas their isomorphism classes are "unlabeled" graphs. Other KS-categories from graph theory are e.g. the category of finite rooted forests or the category of finite directed graphs.

e. Given a group G, let G-$\underline{Sf}$ be the following category: The objects, called finite G-sets, are finite sets on which G acts from the left, and the morphisms are the G-homogeneous maps. For brevity, we will speak only of G-sets when we actually mean finite G-sets.
In G-$\underline{Sf}$, the monomorphisms are the G-homogeneous injective maps and the only initial object is the empty G-set. The direct sum of two G-sets is the disjoint union of the sets with the induced G-operation, and the indecomposable objects are the non-empty G-sets on which G acts transitively. For instance, if V is a subgroup of finite index in G, then the set G/V of left cosets of V becomes an indecomposable G-set by translation. Identifying a subobject $[s:X\to Y]$ of a G-set Y with the G-subset s(X) of Y, a partition of Y in G-$\underline{Sf}$ is

a partition of the set Y whose blocks are G-subsets. The unique KS-partition of a G-set X is the set

$$G\backslash X = \{Gx; x\in X\}$$

of orbits of G in X. Choosing $\underset{\sim}{P}=\{G/V; V\in\underset{\sim}{C}\}$ where $\underset{\sim}{C}$ is a system of representatives of the conjugacy classes of subgroups of finite index in G, the KS-type of a G-set X tells for each V in $\underset{\sim}{C}$ how many orbits of G in X have stabilizers conjugate to V.
Of course, one can define the category G-Sf also for a monoid G. In this case however, it is more difficult to describe the indecomposable objects. As a generalization we consider

(2.5) Functor categories: Given a category $\underline{I}$ with only finitely many objects, let $(\underline{I},\underline{Sf})$ denote the category of all functors from $\underline{I}$ to $\underline{Sf}$. For a functor $Y:\underline{I}\to\underline{Sf}$, the poset Sub(Y) is the Brouwer lattice of all subfunctors of Y ([Joh]). If $\underline{I}$ has no morphisms except the identities, $(\underline{I},\underline{Sf})$ is the product category $\underline{Sf}^I$ where I denotes the finite set of objects in $\underline{I}$. If G is a monoid and if $\underline{I}$ is the category with a single object and morphisms $g\in G$, then $(\underline{I},\underline{Sf})$ is the category G-Sf from above. It is not difficult to verify that $(\underline{I},\underline{Sf})$ is a KS-category with finite automorphism groups. For instance, a functor X can be decomposed as follows: Associate to X the finite undirected graph G(X) whose vertex set is the disjoint union of the sets $X(i), i\in I$, and whose edges are all sets $\{x,y\}$ where $x\in X(i), y\in X(j)$ and $(X\alpha)(x)=y$ for some morphism $\alpha:i\to j$ in $\underline{I}$. Then, for each connected component $C=(V,E)$ of G(X), $X(C):\underline{I}\to\underline{Sf}$, $X(C)(i)=X(i)\cap V$, is an indecomposable subfunctor of X, and

$$\{X(C);\ C \text{ a connected component of } G(X)\}$$

is the unique KS-partition of X in $(\underline{I},\underline{Sf})$.

(2.6) Some notations: Let $N = \{n\in \mathbb{N}_o(P); n\neq 0\}$ and $\Gamma = \mathbb{N}_o(N) = \{\gamma\in \mathbb{N}_o^N; \gamma(n)\neq 0 \text{ for only finitely many } n\in N\}$. For $\gamma\in\Gamma$,

$$|\gamma| = \Sigma_{n\in N}\, \gamma(n) \in \mathbb{N}_o \quad .$$

The weight of γ is

$$\omega(\gamma) = \Sigma_{n\in N}\, \gamma(n)n \in \mathbb{N}_o(P) \quad .$$

Note that always $|\omega(\gamma)| = \Sigma_n \gamma(n)|n| \geq |\gamma| = \Sigma_n \gamma(n)$.

(2.7) DEFINITION: Let $\pi=\{[r:X_i\to Y]; i=1,..,l\}$ be a partition of an object Y in $\underline{K}$. Then the type of π is

$\gamma\in\Gamma$, $\gamma(n)$ = number of blocks of π where X_i has KS-type n .

Observe that $\omega(\gamma)$ is the KS-type of Y and that $|\gamma|=l$ counts the blocks of π.

(2.8) PROPOSITION: Let Y be an object in $\underline{K}$ of KS-type m and let $\gamma\in\Gamma$ have weight m. Then the number of partitions of Y in $\underline{K}$ of type γ is

$$p_\gamma(m) = a(m)/a^\gamma\gamma! \quad .$$

Here $a^\gamma = \Pi_{n\in N}\, a(n)^{\gamma(n)}$ and $\gamma! = \Pi_{n\in N}\, \gamma(n)!$. In particular, the number of Krull-Schmidt partitions of Y in $\underline{K}$ is $a(m)/a^m m!$ where $a^m = \Pi_{P\in \underset{\sim}{P}}\, a(P)^{m(P)}$ and $m! = \Pi_{P\in\underset{\sim}{P}}\, m(P)!$.

(2.9) EXAMPLES:

a. In the category $\underline{Sf}$ of finite sets, we have $N=\{1,2,..\}=\mathbb{N}$ by the convention made in (2.3). The type of a partition π of a finite set Y, defined in (2.7) as

$$\gamma\in\Gamma \quad , \quad \gamma(n) = \text{number of blocks of } \pi \text{ of size } n \ ,$$

is well-known in the literature and often written as $(1^{\gamma(1)}2^{\gamma(2)}..)$ (compare [Be,71]).

b. In the category $\underline{Vf}_{\mathbb{F}}$ of finite-dimensional vector spaces over a finite field $\mathbb{F}$, again $N=\{1,2,..\}$. Given an $\mathbb{F}$-vector space Y of F-dimension m, the type of a partition π of Y is

$$\gamma\in\Gamma \quad , \quad \gamma(n) = \text{number of blocks of } \pi \text{ of } \mathbb{F}\text{-dimension } n \ ,$$

and there are $(q^m-1)..(q^m-q^{m-1})/(q-1)^m m!$ KS-partitions of Y.

Proof of Proposition (2.8): The automorphism group of Y acts on the set of partitions of Y by

$b\pi=\{[br_i:X_i\to Y];i=1,..,l\}$ if $\pi=\{[r_i:X_i\to Y];i=1,..,l\}$,

and leaves the type of partitions unchanged. On the other hand, if $\pi=\{[r_i:X_i\to Y];i=1,..,l\}$, $\rho=\{[s_j:Z_j\to Y];j=1,..,l\}$ are two partitions of Y of the same type, then π,ρ lie in the same orbit of Aut(Y). This is seen as follows: First we can assume that $Z_1=X_1,..,Z_l=X_l$. By the universal property of direct sums, $\oplus_i(X_i,r_i) = Y = \oplus_i(X_i,s_i)$ implies that there exists an automorphism b such that $s_i=br_i$ for i=1,..,l. But then $b\pi=\rho$. We conclude that Aut(Y) acts transitively on the set of partitions of Y of type γ. The order of Aut(Y) is a(m) since Aut(Y) and $\mathrm{Aut}(\oplus_P m(P)P)$ are conjugate by any isomorphism $Y\to\oplus_P m(P)P$, and the stabilizer of a partition of type γ has the order $a^\gamma\gamma!$ according to the following Lemma. So there are $a(m)/a^\gamma\gamma!$ partitions of Y of type γ. □

(2.10) LEMMA: Let Y be an object in $\underline{K}$ of KS-type m and let $\pi=\{[r_i:X_i\to Y];i=1,..,l\}$ be a partition of Y of type γ. Then the stabilizer of π with respect to the action of Aut(Y) on Pa(Y)

$$\{b\in \mathrm{Aut}(Y);b\pi=\pi\}$$

is isomorphic to a direct product of wreath products

$$S = \Pi_{n\in N}\, \mathrm{Aut}(\oplus_P n(P)P) \text{ wr } S_{\gamma(n)}$$

and thus has the order

$$a^{\gamma}\gamma! = \Pi_{n\in N}\ a(n)^{\gamma(n)}\gamma(n)!$$

(Here S_d is the symmetric group of degree $d\geq 0$. Given a finite group G, the wreath product G wr S_d is the set

$$G^d \times S_d = \{(g(1),\dots,g(d);\sigma);\ g(1),\dots,g(d)\in G \text{ and } \sigma\in S_d\}$$

with the multiplication

$$(g(1),\dots,g(d);\sigma)(h(1),\dots,h(d);\tau) = (g(1)h(\sigma^{-1}(1)),\dots,g(d)h(\sigma^{-1}(d));\sigma\tau)$$

and the unit $(1_G,\dots,1_G;id)$. See [JK,81] for details).

Proof: We split π isotypically as $\pi = \dot{\cup}_{n\in N}\ \pi_n$ where $\pi_n = \{[r_n(j):X(n)\to Y];j=1,\dots,\gamma(n)\}$ and $X(n) = \oplus_P n(P)P$. Since $Y = \oplus_{n,j}(X(n),r_n(j))$, we can define a map

$\Psi: S \to Aut(Y)\ ,\ (b_n(1),\dots,b_n(\gamma(n));\sigma_n)_n \to b$,

by the commutative diagrams

$$\begin{array}{ccc} Y & \xrightarrow{\quad b \quad} & Y \\ {\scriptstyle r_n(\sigma_n^{-1}(j))}\Big\uparrow & & \Big\uparrow{\scriptstyle r_n(j)} \\ X(n) & \xrightarrow{\quad b_n(j) \quad} & X(n) \end{array} \qquad (n\in N,\ j=1,\dots,\gamma(n)) .$$

As $br_n(j) = r_n(\sigma_n(j))b_n(\sigma_n(j))$ and hence $[br_n(j)] = [r_n(\sigma_n(j))]$ in Sub(Y), the automorphism b fixes the partition π.

Conversely, if b is an automorphism of Y in the stabilizer of π, we have for every $n\in N$ a permutation $\sigma_n\in S_{\gamma(n)}$ such that $[br_n(j)] = [r_n(\sigma_n(j))]$ for $j=1,\dots,\gamma(n)$. Moreover, there are automorphisms $b_n(1),\dots,b_n(\gamma(n))$ of X(n) such that $br_n(j) = r_n(\sigma_n(j))b_n(\sigma_n(j))$ for $j=1,\dots,\gamma(n)$. Thus $\Psi((b_n(1),\dots,b_n(\gamma(n));\sigma_n)_n) = b$, proving that the image of Ψ is the stabilizer of π. Ψ is one-to-one because the σ_n and the $b_n(j)$ are uniquely determined by b. For this, observe that the $r_n(j)$ are monomorphisms. Finally, we see from the commuting diagrams

$$\begin{array}{ccccc} Y & \xrightarrow{\qquad c \qquad} & Y & \xrightarrow{\qquad b \qquad} & Y \\ {\scriptstyle r_n((\sigma_n\tau_n)^{-1}(j))}\Big\uparrow & & \Big\uparrow{\scriptstyle r_n(\sigma_n^{-1}(j))} & & \Big\uparrow{\scriptstyle r_n(j)} \\ X(n) & \xrightarrow[c_n(\sigma_n^{-1}(j))]{} & X(n) & \xrightarrow[b_n(j)]{} & X(n) \end{array}$$

that Ψ is a group homomorphism. □

(2.11) LEMMA: Let A be a commutative algebra (with unit 1) over the rational numbers, and let $c:N\to A$ be a function. Then

$$\exp(\Sigma_{n\in N}\ c(n)\ z^n/a(n)) = 1 + \Sigma_{m\in N}\ (\Sigma_{\gamma\in\Gamma,\omega(\gamma)=m}\ p_\gamma(m)c^\gamma)\ z^m/a(m)$$

in the power series algebra $A[[z(P);P\in\underset{\sim}{P}]]$. Here $c^\gamma = \Pi_{n\in N}\ c(n)^{\gamma(n)}$, and $p_\gamma(m)$ is the number of partitions of $\oplus_P m(P)P$ of type γ .

Proof: By (A.10) from the appendix, $A[[z(P);P\in \underset{\sim}{P}]]$ is an abstract incidence algebra over A. In order to apply Proposition (A.14), we review the situation of (A.14) in this special case:
The topological basis $(e(t);t\in T)$ is $(z^n;n\in \mathbb{N}_o(\underset{\sim}{P}))$ where $z^n = \Pi_P z(P)^{n(P)}$. Thus $T=\mathbb{N}_o(\underset{\sim}{P})$, $T(1)=N$ and $\mathbb{N}_o(T(1))=\Gamma$. Since the multiplication constants from (A.8) are
$G(n;n_1,..,n_l)=1$ if $n=n_1+..+n_l$ and $=0$ else, we have
$G(n;\gamma)=|\gamma|!/\gamma!$ if $\omega(\gamma)=n$ and $=0$ otherwise.
Hence $\exp(\Sigma_n c(n)z^n/a(n)) = 1+\Sigma_m(\Sigma_{\gamma,\omega(\gamma)=m} c^\gamma/a^\gamma\gamma!)z^m$ by (A.14), and the result follows from Proposition (2.8). □

(2.12) DEFINITION: Let $A=\mathbb{Q}[V,W_n;n\in N]$ be the polynomial algebra in the variables V and W_n, $n\in N$. We define the <u>partial Bell polynomials</u> in <u>K</u> for all $m\in N$ and $l=1,..,|m|$ by

$$B_{m,l}(W) = \Sigma\, p_\gamma(m) W^\gamma \qquad (\; W^\gamma = \Pi_n\, W_n^{\gamma(n)}\;)$$

where the sum runs over all $\gamma\in\Gamma$ with $\omega(\gamma)=m$ and $|\gamma|=l$. We call

$$Y_m(W) = \Sigma_{l=1}^{|m|} B_{m,l}(W)$$

the <u>complete Bell polynomials</u> in <u>K</u>. By the definition,

$B_{m,l}(1,1,..)$ = number of partitions of $\oplus_P m(P)P$ with l blocks

and

$Y_m(1,1,..)$ = number of all partitions of $\oplus_P m(P)P$.

Of course, this is a generalization of the classical Bell polynomials ([Co,74]) which are obtained when <u>K</u> is the category of finite sets. Choosing $c:N\to A$, $c(n)=VW_n$, in Lemma (2.11), we get the following identity.

(2.13) THEOREM: (the <u>exponential formula for partitions</u>)
In the power series algebra $A[[z(P);P\in\underset{\sim}{P}]]$ where $A=\mathbb{Q}[V,W_n;n\in N]$ is the polynomial algebra,

$$\exp(V\,\Sigma_{n\in N} W_n\, z^n/a(n)) = 1 + \Sigma_{m\in N}\,(\Sigma_{l=1}^{|m|} B_{m,l}(W)\, V^l)\, z^m/a(m) \quad ,$$

in particular

$$\exp(\Sigma_{n\in N} W_n\, z^n/a(n)) = 1 + \Sigma_{m\in N}\, Y_m(W)\, z^m/a(m) \quad . \quad \square$$

(2.14) REMARK: A comparable result in a different situation was derived by R.P.Stanley in [St,78']. In the examples where $\underset{\sim}{P}$ has only one member P (and so $\mathbb{N}_o(\underset{\sim}{P})=\mathbb{N}_o$, according to our convention), his number M(n) coincides with the number of KS-partitions of $nP=P\oplus..\oplus P$ (n times) which is $a(n)/a(1)^n n!$ by Proposition (2.8), for every n.

(2.15) EXAMPLES:

a. In the category $\underline{Sf}$ of finite sets, Theorem (2.13) yields the generating function of the Stirling numbers of the second kind

$$\exp(V(e^z-1)) = 1+ \Sigma_{m=1}^{\infty} \; (\Sigma_{l=1}^{m} \; S(m,l)V^l) \; z^m/m!$$

and the generating function of the Bell numbers

$$\exp(e^z-1) = 1+ \Sigma_{m=1}^{\infty} \; B(m)z^m/m!$$

([Co,74]).

b. In the category $\underline{Vf}_{\mathbb{F}}$ of finite-dimensional vector spaces over a finite field $\mathbb{F}$ with q elements, let

$$B_q(m) = \text{number of all partitions of } \mathbb{F}^m \; .$$

Then, by (2.13), $\exp(\Sigma_{n=1}^{\infty} z^n/b_q(n)) = 1+ \Sigma_{m=1}^{\infty} B_q(m)z^m/b_q(m)$ where $b_q(n)=(q^n-1)(q^n-q)..(q^n-q^{n-1})$ is the order of the general linear group in n dimensions. Replacing z by (q-1)z and using the q-factorials $[n]!=[n][n-1]..[1]$, $[n]=(q^n-1)/(q-1)$, we get the formula

$$\exp(\Sigma_{n=1}^{\infty} \; z^n/q^{n(n-1)/2}[n]!) = 1+ \Sigma_{m=1}^{\infty} \; B_q(m)z^m/q^{m(m-1)/2}[m]!$$

which converges to the corresponding identity in sets as q tends to 1 ([BG,71],Example 11).

(2.16) REMARK: If $b_1:X_1 \to X_1$, $b_2:X_2 \to X_2$ are automorphisms in $\underline{K}$ and if X is the direct sum of X_1, X_2 with the canonical injections $u_1:X_1 \to X$, $u_2:X_2 \to X$, we can define an automorphism $b_1 \oplus b_2$ of X by the commutative diagrams

$$\begin{array}{ccc} X_i & \xrightarrow{\;b_i\;} & X_i \\ {\scriptstyle u_i}\downarrow & & \downarrow{\scriptstyle u_i} \\ X & \xrightarrow{\;b_1 \oplus b_2\;} & X \end{array} \qquad (i=1,2) \; .$$

Hence the direct product $Aut(X_1) \times Aut(X_2)$ is isomorphic to a subgroup of $Aut(X)$ by the map $(b_1,b_2) \to b_1 \oplus b_2$. In particular, $a(m)a(n)$ divides $a(m+n)$ for all $m,n \in \mathbb{N}_o(\underset{\sim}{P})$.

We conjectured in (1.49),g that generating functions $\Sigma_n f(n)z^n/\varphi(n)$ arising in enumeration problems have the following property:

$$\varphi(m+n)/\varphi(m)\varphi(n) \in \mathbb{N} \qquad \text{for all } m,n \in \mathbb{N}_o(I) \; .$$

In this book, φ often will be given by the order of the automorphism groups in a KS-category and thus has this property.

(2.17) Krull-Schmidt categories with unique KS-partitions:

We finally examine the case that every object in $\underline{K}$ has a unique KS-partition. Let Y be an object in $\underline{K}$ of KS-type m. From Lemma (2.10) we see that the automorphism group of Y is isomorphic to a direct product of wreath products

$$\Pi_{P\in \underset{\sim}{P}} \mathrm{Aut}(P) \text{ wr } S_{m(P)}$$

and has the order

(2.18) $$a(m) = \Pi_{P\in \underset{\sim}{P}} a(\varepsilon(P))^{m(P)} m(P)! = a^m m! \quad .$$

Substituting this into Proposition (2.8), we get a formula for the number of partitions of Y in $\underline{K}$ of type γ, $\omega(\gamma)=m$, which is nearly independent of the category $\underline{K}$: $p_\gamma(m) = m!/\Pi_{n\in N}(n!)^{\gamma(n)}\gamma(n)!$.

(2.19) EXAMPLE: Let G be a group. In the category G-$\underline{Sf}$ of finite G-sets, introduced in (2.4),e, KS-partitions are unique and $\underset{\sim}{P}=\{G/V; V\in \underset{\sim}{C}\}$ where $\underset{\sim}{C}$ is a system of representatives of the conjugacy classes of subgroups of finite index in G. As the automorphism group of the G-set G/V is anti-isomorphic to the factor group $N_G(V)/V$ where $N_G(V)$ denotes the normalizer of V in G, $a(\varepsilon(G/V))$ is the index of V in $N_G(V)$, and

$$a(m) = \Pi_{V\in \underset{\sim}{C}}[N_G(V):V]^{m(G/V)} m(G/V)! \quad .$$

§2. Functors

In this section we consider two Krull-Schmidt categories $\underline{H},\underline{K}$ with finite automorphism groups and unique KS-partitions, and a faithful functor $F:\underline{H}\to\underline{K}$ with the properties (F1) and (F2) below. If X is an object in $\underline{K}$ and if D is an object in $\underline{H}$ such that FD=X, we call D an <u>F-structure</u> on X.

(F1) For every object X in $\underline{K}$, there are only finitely many F-structures on X.

We denote the finite set of F-structures on X by $F^{-1}(X)$.

(F2) F-structures on direct summands induce a unique F-structure on the direct sum:
If $X=\oplus_{j\in J}(X_j,u_j)$ in $\underline{K}$ where J is a finite index set, and if D_j is an F-structure on X_j for each j, then there exists a unique F-structure D on X and monomorphisms $\hat{u}_j:D_j\to D$ such that $F\hat{u}_j=u_j$ for all j and $D=\oplus_{j\in J}(D_j,\hat{u}_j)$.

(2.20) LEMMA:

(i) For any initial object Z in $\underline{K}$, there is exactly one F-structure on Z.

(ii) F preserves partitions: If $\varepsilon=\{[v_j:D_j\to E];j=1,..,l\}$ is a partition of an object E in $\underline{H}$, then $F\varepsilon=\{[Fv_j:FD_j\to FE];j=1,..,l\}$ is a partition of the object FE in $\underline{K}$.

Proof: (i) Applying (F2) when J is empty, we see that there exists a unique initial F-structure D on Z. But any F-structure E on Z must be initial in $\underline{H}$ because otherwise the canonical injections $u_1,u_2:E\to E\oplus E$ would be different and hence $Fu_1,Fu_2:Z\to F(E\oplus E)$ too.
(ii) Let X be the direct sum of the FD_j in $\underline{K}$ with the canonical injections u_j. By (F2) there exist an object D and monomorphisms $\hat{u}_j:D_j\to D$ in $\underline{H}$ such that $D=\oplus_j(D_j,\hat{u}_j)$. Since also $E=\oplus_j(D_j,v_j)$, there is an isomorphism $b:D\to E$ in $\underline{H}$ such that $b\hat{u}_j=v_j$ for all j. Hence $(Fb)u_j=Fv_j$ for all j implying $FE=\oplus_j(FD_j,Fv_j)$ where no FD_j is initial. □

Now fix a system $\underset{\sim}{P}$ of representatives of the isomorphism classes of indecomposable objects in $\underline{K}$, and let N and Γ be defined as in (2.6).
(2.21) For any Krull-Schmidt type $n\in N$, we denote the number of indecomposable F-structures on $\oplus_P n(P)P$ by $f(n)$. Thus we have a map

$$f:\ N\to \mathbb{N}_o\ .$$

(2.22) DEFINITION: If E is an object in $\underline{H}$ with the KS-partition ε, we call the type $\gamma\in\Gamma$ of the partition Fε in $\underline{K}$ the <u>underlying type</u> of E.

(2.23) PROPOSITION: Let Y be an object in $\underline{K}$ of KS-type m, and let $\gamma\in\Gamma$ have weight m. Then the number of F-structures E on Y of underlying type γ is

$$g_\gamma(m) = p_\gamma(m)f^\gamma \qquad (f^\gamma=\Pi_{n\in N} f(n)^{\gamma(n)})$$

where $p_\gamma(m)$ denotes the number of partitions of Y in $\underline{K}$ of type γ.

Proof: Let $\pi=\{[r_j:X_j\to Y];j=1,..,l\}$ be a partition of Y of type γ. Applying (F2) to $Y=\oplus_j(X_j,r_j)$, we can define a map

$$\Phi:\ \Pi_{j=1}^{l}(F^{-1}(X_j)\cap\mathrm{Ind}) \to \{\varepsilon;\varepsilon \text{ a KS-partition in } \underline{H} \text{ with } F\varepsilon=\pi\}$$
$$(D_j) \to \{[\hat{f}_j:D_j\to D];j=1,..,l\}$$

which will turn out to be a bijection. Here Ind is the class of indecomposable objects in $\underline{H}$.
To prove that Φ is one-to-one, suppose that $\{[\hat{f}_j:D_j\to D];j=1,..,l\}=$ $=\{[\hat{f}'_j:D'_j\to D];j=1,..,l\}$. Then there exist a permutation $\sigma\in S_l$ and isomorphisms $d_j:D_j\to D'_{\sigma(j)}$ such that $\hat{f}_j=\hat{f}'_{\sigma(j)}d_j$ for $j=1,..,l$. Since $F\hat{f}_j=r_j=F\hat{f}'_j$, we find that $r_j=r_{\sigma(j)}(Fd_j)$ for all j. But direct sums in $\underline{K}$ are disjoint and the r_j are monomorphisms. Hence σ=id and each Fd_j is the identity on X_j. By the following Lemma, also the d_j are identities. Thus $D_j=D'_j$ for all j as desired.

(2.24) LEMMA: Every isomorphism in $\underline{H}$ mapped by F to an identity in $\underline{K}$, is an identity in $\underline{H}$.

Proof: Let $d:B\to C$ be an isomorphism in $\underline{H}$ such that Fd is the identity on $X=FB=FC$. We apply (F2) to $X=\oplus(X,id)$ (one summand) and to the F-structure B on X. Since both $B,id:B\to B$ and $C,d:B\to C$ satisfy the conditions in (F2), we have $B=C$ and $d=id$ by uniqueness. □

We continue the proof of Proposition (2.23) by showing that Φ is onto. Let ε be a KS-partition in $\underline{H}$ with $F\varepsilon=\pi$. We can assume that $\varepsilon=\{[v_j:E_j\to E];j=1,..,l\}$ where $[Fv_j]=[r_j]$ for all j, otherwise relabel the subobjects. Then there are isomorphisms $b_j:FE_j\to X_j$ such that $Fv_j=r_jb_j$, $j=1,..,l$. For each j, $X_j=\oplus(FE_j,b_j)$ (one summand), and by (F2) there exist an object D_j and a monomorphism $\hat{b}_j:E_j\to D_j$ in $\underline{H}$ such that $FD_j=X_j$, $F\hat{b}_j=b_j$ and $D_j=\oplus(E_j,\hat{b}_j)$ (one summand). But here $\hat{b}_j$ must be an isomorphism, and $D_j\in\mathrm{Ind}$ because E_j is indecomposable. Now define monomorphisms in $\underline{H}$ by $w_j=v_j(\hat{b}_j)^{-1}$, $j=1,..,l$. Then $\varepsilon=\{[w_j:D_j\to E];j=1,..,l\}$, $Fw_j=(Fv_j)(b_j)^{-1}=r_j$ for all j, and $E=\oplus_j(D_j,w_j)$. Looking at (F2) we see that $D=E$ and $\hat{r}_j=w_j$ for all j. Thus $\varepsilon=\Phi((D_j))$, and Φ is a bijection.
If X_j has the KS-type n, then $F^{-1}(X_j)\cap\mathrm{Ind}$ has $f(n)$ elements by (2.21). So the number of KS-partitions in $\underline{H}$ with $F\varepsilon=\pi$ is $\Pi_{n\in N}f(n)^{\gamma(n)}=f^{\gamma}$ which ends the proof. □

(2.25) THEOREM: (the <u>exponential formula for functors</u>)
In the power series algebra $A[[z(P);P\in\underset{\sim}{P}]]$ where $A=\mathbb{Q}[V,W_n;n\in N]$ is the polynomial algebra,

$$\exp(V\,\Sigma_{n\in N}\,f(n)W_n\,z^n/n!) = 1+\Sigma_{m\in N}\,(\Sigma_{l=1}^{|m|}\,B_{m,l}(fW)\,V^l)\,z^m/m! \quad .$$

Here the $B_{m,l}(W)$ are the partial Bell polynomials in $\underline{K}$ defined in (2.12), and $B_{m,l}(fW)=B_{m,l}(f(n)W_n;n\in N)=\Sigma\,g_\gamma(m)W^\gamma$, the summation being taken over all types $\gamma\in\Gamma$ with $\omega(\gamma)=m$ and $|\gamma|=l$. In particular,

$$\exp(\Sigma_{n\in N}\,f(n)W_n\,z^n/n!) = 1+\Sigma_{m\in N}\,Y_m(fW)\,z^m/m!$$

where the $Y_m(W)$ are the complete Bell-polynomials in $\underline{K}$ and $Y_m(fW)=Y_m(f(n)W_n;n\in N)=\Sigma_{\gamma,\omega(\gamma)=m}\,g_\gamma(m)W^\gamma$.

Proof: We apply Lemma (2.11) to $c:N\to A$, $c(n)=f(n)VW_n$, and obtain
$$\exp(V\Sigma_{n\in N}f(n)W_nz^n/a(n)) = 1+\Sigma_{m\in N}(\Sigma_{l=1}^{|m|}\,B_{m,l}(fW)V^n)\,z^m/a(m) \quad .$$
Since KS-partitions in $\underline{K}$ are unique, $a(n)=a^n n!$ by (2.18). Replacing each variable $z(P)$ by $a(\varepsilon(P))z(P)$, we can cancel out a^n. □

(2.26) COROLLARY: Let $g_l(m)$ denote the number of F-structures with l components on an object of KS-type m. Then

$$\exp(V\,\Sigma_{n\in N}\ f(n)\ z^n/n!) = 1 + \Sigma_{m\in N}\ (\Sigma_{l=1}^{|m|}\ g_l(m)V^l)\ z^m/m! \quad .$$

The generating function for the number g(m) of all F-structures on an object of KS-type m is

$$\exp(\Sigma_{n\in N}\ f(n)\ z^n/n!) = 1 + \Sigma_{m\in N}\ g(m)\ z^m/m! \quad . \quad \square$$

(2.27) REMARK: It is worth noting that here (multivariate) exponential power series appear because the automorphism groups of objects in K are wreath products. In the literature, the connection between the number of connected and the number of arbitrary structures by exponential formulas has been studied from various viewpoints. D.Foata and M.P.Schützenberger ([FS,70],[F,74]) use the concept of a "composé partitionell" whereas E.A.Bender and J.R.Goldman ([BG,71]) operate with "prefabs". Our approach is most closely to that of R.P.Stanley ([St,78], [St,78']) by "exponential structures", in particular "r-partitions", and that of A.Joyal ([Joy,81]) by "espèces".

(2.28) EXAMPLES: In the rest of this section, we briefly look at some examples of the literature from our point of view. In each case the required conditions are immediately verified. A new application of the exponential formula to the enumeration of quotient objects is given in the next chapter.

a. Labeled graphs:

Let K be the category Sf of finite sets, let H be the category Gf of (finite undirected simple) graphs and define a faithful functor F:H→K by mapping a graph (V,E) to its vertex set V and a graph homomorphism s:(V,E)→(W,F) to the function s:V→W. Here N={1,2,3,..}= ℕ, f(n) counts the connected graphs on the vertex set {1,..,n}, and $g(m)=2^{m(m-1)/2}$ is the total number of graphs on {1,..,m}. By Corollary (2.26),

$$\exp(\Sigma_{n=1}^{\infty}\ f(n)\ z^n/n!) = 1 + \Sigma_{m=1}^{\infty}\ 2^{m(m-1)/2}\ z^m/m! \quad .$$

Of course, instead of Gf also the category of k-colorable graphs or the category of finite rooted forests can be chosen, thereby obtaining similar identities ([BG,71],Examples 6 and 8).

b. Group actions on finite sets:

Let G be a monoid such that, for every finite set X, G acts on X in only finitely many ways (e.g. if G is finitely generated). Let H be the category G-Sf of finite G-sets, and define a faithful functor F:H→Sf by mapping a G-set to its underlying set and a G-homogeneous map to itself. Here $N=\{1,2,3,..\}$, $f(n)$ is the number of indecomposable operations of G on $\{1,..,n\}$ whereas $g(m)$ counts all G-actions on $\{1,..,m\}$. Hence, by (2.26),

$$\exp(\Sigma_{n=1}^{\infty} f(n)\ z^n/n!) = 1 + \Sigma_{m=1}^{\infty} g(m)\ z^m/m! \quad .$$

For instance, if G is the monoid generated by one element subject to the relation $x^2=x$, then $f(n)=n$ and

$$1 + \Sigma_{m=1}^{\infty} g(m)\ z^m/m! = \exp(ze^z)$$

is the generating function of the number $g(m)$ of idempotent functions from $\{1,..,m\}$ into itself ([Co,74],p.91).

Now let G be a group. Then the condition for G stated above can be formulated as follows: For every positive integer n, there are only finitely many subgroups of G of index n.

We denote this number by $s_n(G)$. Given a subgroup H of G of index n, the transitive operations of G on $\{1,..,n\}$ such that G is the stabilizer of 1 are in one-to-one correspondence to the bijections from G/H to $\{1,..,n\}$ mapping the coset H to 1, a transitive operation corresponding to its orbit map through 1. Therefore, $f(n)=(n-1)!s_n(G)$ and

$$1 + \Sigma_{m=1}^{\infty} g(m)\ z^m/m! = \exp(\Sigma_{n=1}^{\infty} s_n(G)\ z^n/n) \quad .$$

In the literature (e.g. in [St,78]), special cases are considered: If G is the cyclic group of order d, then $g(m)$ counts the solutions of the equation $x^d=1$ in the symmetric group S_m, and we get the formula of Chowla, Herstein and Scott ([CHS,52])

$$1 + \Sigma_{m=1}^{\infty} g(m)\ z^m/m! = \exp(\Sigma_{n \text{ divides } d}\ z^n/n) \quad .$$

If $G=F_r$ is the free group on r generators, then

$$\Sigma_{n=1}^{\infty} s_n(F_r)\ z^n/n = \log(1+ \Sigma_{m=1}^{\infty}(m!)^{r-1}z^m) \quad .$$

Finally, if G is the infinite cyclic group, then $f(n)=(n-1)!$ for all n, $Y_m(fW)/m!$ is the cycle indicator polynomial of the symmetric group S_m, and Theorem (2.25) yields the well-known identity

$$1 + \Sigma_{m=1}^{\infty} Y_m(fW)/m!\ z^m = \exp(\Sigma_{n=1}^{\infty} W_n\ z^n/n) \quad .$$

c. <u>Roots in the symmetric group:</u>

Let $\underline{K}$ be the following category: The objects are pairs (X,σ) where X is a finite set and σ is a permutation of X. A morphism s from (X,σ) to (Y,τ) is a map $s:X\to Y$ such that $s\sigma=\tau s$.

Obviously $\underline{K}$ is isomorphic to the category $\mathbb{Z}$-$\underline{Sf}$ of finite sets on which the infinite cyclic group $\mathbb{Z}$ acts from the left. Since the indecomposable objects in $\underline{K}$ are finite sets with a cycle, we can choose

$$\underset{\sim}{P} = \{(\{1,..,i\},(12..i));i=1,2,..\}$$

and identify the KS-type of (X,σ) with the cycle type of σ

$$\alpha = (1^{\alpha(1)}2^{\alpha(2)}..) \quad , \quad \alpha(i) = \text{number of cycles of } \sigma \text{ of length } i \quad .$$

Hence $N=\{\alpha\in \mathbb{N}_o(\mathbb{N});\alpha\neq 0\}$.

Now let k be a positive integer. Put $\underline{H}=\underline{K}$ and define a faithful functor $F:\underline{H}\to\underline{K}$ by $F(X,\sigma)=(X,\sigma^k)$ and $Fs=s$. If σ is a permutation of $\{1,..,l\}$ of cycle type α, then $g(\alpha)$ equals the number $r_k(\alpha)$ of all k-th roots of σ in S_l whereas $f(\alpha)$ counts cyclic roots. Given a positive integer l, it is not difficult to see that the only non-zero value $f(\alpha)$ where $\alpha\in N$ satisfies $\Sigma_{i=1}^{\infty} i\alpha(i)=l$, is

$$f(j\varepsilon(l/j)) = (l/j)^{j-1}(j-1)! \quad .$$

Here $j=(l,k)$ is the greatest common divisor of l and k, and $\varepsilon(i)$ is the i-th standard basis vector of $\mathbb{N}_0(\mathbb{N})$. Thus, by Corollary (2.26),

$$1+\Sigma_{\alpha\in N} r_k(\alpha)z^\alpha/\alpha! = \exp(\Sigma_{l=1}^{\infty}(l/j)^{j-1}(j-1)!(z_{l/j})^j/j!) \quad \text{in } \mathbb{Q}[[z_1,z_2,..]]$$

$(z^\alpha = \Pi_{i=1}^{\infty}(z_i)^{\alpha(i)})$.

Rearranging the exponent gives the following result:

$$1 + \Sigma_{\alpha\in N} r_k(\alpha) z^\alpha/\alpha! = \exp(\Sigma_{i=1}^{\infty} \Sigma_{d\in T(k;i)} (iz_i)^d/id)$$

where $T(k;i)$ is the set of all (positive) divisors d of k such that k/d and i are relatively prime. By differentiating we obtain a recursion formula for the $r_k(\alpha)$:

$$r_k(0)=1 \qquad \text{and} \qquad r_k(\alpha+\varepsilon(i)) = \Sigma\ (\alpha(i))_{d-1}\, i^{d-1}\, r_k(\alpha-(d-1)\varepsilon(i))$$

where $(\alpha(i))_{d-1}=\alpha(i)(\alpha(i)-1)..(\alpha(i)-d+2)$ is a falling factorial and the sum runs over all divisors d of k such that $(k/d,i)=1$.

For instance, consider the permutation (12)(34) in S_4 which has the two quadratic roots (1324) and (1423). Here $r_2(2\varepsilon(2)) = 1\cdot 2^1\cdot r_2(0) = 2$.

This recursion has been derived directly in [KT,83],vol.II,pp.227.

An explicit, but complicated formula for $r_k(\alpha)$ is given in [Pa,81].

CHAPTER III

SHEAFLIKE CATEGORICAL STRUCTURES AND INCIDENCE BIALGEBRAS

In chapter I we have constructed the incidence algebra of a categorical structure $\underline{K},M,\sim$. In this chapter we introduce those "sheaflike" categorical structures for which the incidence algebra becomes a cocommutative topological bialgebra.

§1. Sheaflike categorical structures

Let $\underline{K}$ be a Krull-Schmidt category with unique KS-partitions.
We denote the category of morphisms in $\underline{K}$ by Mor($\underline{K}$). Its objects are the morphisms in $\underline{K}$. Given two objects r,s in Mor($\underline{K}$), the morphisms in Mor($\underline{K}$) from r to s are all pairs (c,d) of morphisms in $\underline{K}$ which make the diagram

$$\begin{array}{ccc} X & \xrightarrow{r} & Y \\ {\scriptstyle c}\downarrow & & \downarrow{\scriptstyle d} \\ Z & \xrightarrow{s} & W \end{array}$$

commute. Then Mor($\underline{K}$) is a skeletal-small category which inherits from $\underline{K}$ the properties ($\underline{K}$1)-($\underline{K}$3):
If 0 is an initial object in $\underline{K}$, then $0\to 0$ is an initial object in Mor($\underline{K}$). The direct sum of two objects $s_1:X_1\to Y_1$ and $s_2:X_2\to Y_2$ in Mor($\underline{K}$) exists and is selected as follows: If X is the direct sum of X_1,X_2 with the canonical injections u_1,u_2 and if Y is the direct sum of Y_1,Y_2 with the canonical injections v_1,v_2, then $s_1\oplus s_2$ is the uniquely determined morphism in $\underline{K}$ such that the diagrams

$$\begin{array}{ccc} X_i & \xrightarrow{s_i} & Y_i \\ {\scriptstyle u_i}\downarrow & & \downarrow{\scriptstyle v_i} \\ X & \xrightarrow{s_1\oplus s_2} & Y \end{array} \qquad (i=1,2)$$

commute. Here the canonical injections $(u_1,v_1):s_1\to s_1\oplus s_2$ and $(u_2,v_2):s_2\to s_1\oplus s_2$ are monomorphisms. In fact, a morphism (c,d) in Mor($\underline{K}$)

is a monomorphism if and only if the morphisms c,d in $\underline{K}$ are monomorphisms (owing to the existence of initial objects in $\underline{K}$). To avoid confusion we call subobjects in Mor($\underline{K}$) submorphisms. Since any submorphism $[(c,d):r\to s]$ of s, where $r:X\to Y$ and $s:Z\to W$, gives rise to subobjects $[c:X\to Z]$ of Z and $[d:Y\to W]$ of W, direct sums in Mor($\underline{K}$) are disjoint.

Let M be a class of morphisms in $\underline{K}$ with the properties (M1)-(M3). We assume in addition that M satisfies the conditions (M4)-(M6) below:

(M4) For any morphisms s_1,s_2 in $\underline{K}$, $s_1\oplus s_2\in M$ if and only if $s_1,s_2\in M$.

(M5) If the codomain of a morphism s in M is initial, then also s is initial.

(3.1) DEFINITION: Let $s:X\to Y$ be a morphism in M and let $\sigma=\{[(c_i,d_i):s_i\to s];i=1,..,l\}$ be a partition of s in Mor($\underline{K}$). Then, by (M5),

$\pi = \{[d_i:Y_i\to Y];i=1,..,l\}$

is a partition of Y in $\underline{K}$ which we call the image partition of Y induced by σ.

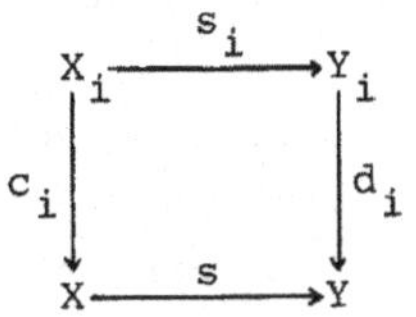

The most important condition for M is

(M6) For any morphism $s:X\to Y$ in M and any partition $\pi=\{[r_1:Y_1\to Y],[r_2:Y_2\to Y]\}$ of Y, there exists a unique partition σ of s such that π is the image partition of Y induced by σ.

By induction (M5) and (M6) imply

(3.2) LEMMA: Let $s:X\to Y$ be a morphism in M. Then the map $\Phi:\mathrm{Pa}(s)\to\mathrm{Pa}(Y)$ sending a partition of s to the induced image partition of Y, is a bijection. Hence s is indecomposable if and only if Y is indecomposable. If π is the unique Krull-Schmidt partition of Y in $\underline{K}$, then $\Phi^{-1}(\pi)$ is the unique KS-partition of s in Mor($\underline{K}$). □

(3.3) EXAMPLES: In the category $\underline{Sf}$ of finite sets, (M6) holds for all morphisms: If $s:X\to Y$ is a map and if $\pi=\{Y_1,Y_2\}$ is a partition of Y, then the unique partition σ of s is $\{s_1:s^{-1}(Y_1)\to Y_1, s_2:s^{-1}(Y_2)\to Y_2\}$ where the s_i are the restrictions of s. The unique KS-partition of s is $\{s_y:s^{-1}(y)\to\{y\};y\in Y\}$.

(If in a category $\underline{K}$ subobjects are identified with objects by distinguished representatives, then submorphisms can be identified with morphisms in K in the same way: A submorphism $[(c,d):r\to s]$ of $s:X\to Y$ is identified with the uniquely determined morphism s' in K such that the diagram

$$\begin{array}{ccc} X & \xrightarrow{s} & Y \\ \uparrow & & \uparrow \\ X' & \xrightarrow{s'} & Y' \end{array}$$

commutes. Here $X'\to X$ and $Y'\to Y$ are distinguished representatives of c or d respectively. For instance, in the categories of sets, groups or modules, the distinguished representatives are the inclusion maps and subobjects correspond to subsets, subgroups or submodules resp.)
In the categories of finite graphs, finite rooted forests or finite sets under group action, too, all morphisms fulfill (M6). In the category of finite-dimensional vector spaces over a finite field however, the class of monomorphisms does not satisfy (M6) as the inclusion $X\to Y=Y_1\oplus Y_2$ in general is not the direct sum of the inclusions $X\cap Y_1\to Y_1$ and $X\cap Y_2\to Y_2$.

(3.4) Let H denote the full subcategory of Mor(K) whose objects are the morphisms in M. Then, by (M4), direct sums in H exist and can be chosen as in Mor(K). From Lemma (3.2) we see that H is a Krull-Schmidt category with unique KS-partitions.

(3.5) LEMMA: If $s_1:X_1\to Y_1$ and $s_2:X_2\to Y_2$ are morphisms in K such that $s_1\oplus s_2:X_1\oplus X_2\to Y_1\oplus Y_2$ is an isomorphism, then s_1 and s_2 are isomorphisms. Therefore, in every Krull-Schmidt decomposition of an isomorphism in K all direct summands are isomorphisms.

Proof: Set $X=X_1\oplus X_2$, $Y=Y_1\oplus Y_2$ and $s=s_1\oplus s_2$. We apply (M6) to the identity id_Y and to the partition $\{[Y_1\to Y],[Y_2\to Y]\}$ of Y. By the uniqueness assertion in (M6) it follows from the diagrams

$$\begin{array}{ccc} Y_i & \xrightarrow{id_{Y_i}} & Y_i \\ \downarrow & & \downarrow \\ Y & \xrightarrow{id_Y} & Y \end{array} \qquad \text{and} \qquad \begin{array}{cccc} & X_i & \xrightarrow{s_i} & Y_i \\ X \swarrow & & & \downarrow \\ \searrow^{s} & Y & \xrightarrow{id_Y} & Y \end{array} \qquad (i=1,2)$$

that $[id_{Y_i}\to id_Y]=[s_i\to id_Y]$ in $Sub(id_Y)$, and so the s_i are isomorphisms. □

Finally, let ~ be an equivalence relation on M with the properties (~1)-(~4). Moreover we assume that ~ has the properties (~5) and (~6) stated below.

(~5) If $r_1\sim r_2$ and $s_1\sim s_2$ in M, then also $r_1\oplus s_1\sim r_2\oplus s_2$.

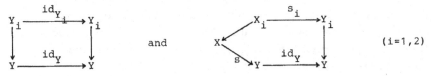

Thus the set of types $T = M/\sim = \{\bar{s}; s \in M\}$ becomes a commutative monoid with addition $\bar{r} + \bar{s} = \overline{r \oplus s}$ and zero element $0 = \overline{0 \to 0}$.

(3.6) DEFINITION: A type t is called <u>indecomposable</u> if $t \neq 0$ and if $t = t_1 + t_2$ for some types t_1, t_2 implies that $t_1 = 0$ or $t_2 = 0$. Let U denote the set of indecomposable types.

The last condition for $\sim$ is

($\sim$6) Every type is the sum of indecomposable types, and this representation is unique up to the order.

Hence the commutative monoid T is free with basis U, i.e.

(3.7) $$T = \oplus_{u \in U} \mathbb{N}_o u \quad , \quad t = \Sigma_{u \in U}\, t(u) u \quad .$$

By ($\sim$2) we can decompose U into the disjoint subsets
$U' = \{\bar{s} \in U;\ s \text{ an isomorphism}\}$ and $U'' = \{\bar{s} \in U;\ s \text{ no isomorphism}\}$.
Lemma (3.5) implies that

(3.8) $$T' = \oplus_{u \in U'} \mathbb{N}_o u$$

is a free submonoid of T with basis U'. Putting

(3.9) $$T'' = \oplus_{u \in U''} \mathbb{N}_o u$$

we obtain the direct sum decomposition of abelian monoids $T = T' \oplus T''$.

(3.10) The <u>standard example</u> of equivalence is the isomorphism relation in Mor($\underline{K}$) which always satisfies ($\sim$1)-($\sim$6). The conditions ($\sim$1)-($\sim$4) have been checked in (1.5), and ($\sim$5) is obvious. For ($\sim$6) observe that the indecomposable types are just the isomorphism types of indecomposable morphisms in M, and that every morphism in M has a unique KS-partition in Mor($\underline{K}$).

(3.11) DEFINITION: We call a triple $\underline{K}, M, \sim$ where $\underline{K}$ is a Krull-Schmidt category with unique KS-partitions, M is a class of morphisms in $\underline{K}$ with (M1)-(M6) and $\sim$ is an equivalence relation on M with ($\sim$1)-($\sim$6), a <u>sheaflike categorical structure</u>.

For instance, toposes (i.e. categories of sheaves, [GV,72]) satisfying the required finiteness conditions give rise to sheaflike categorical structures, in particular the functor categories introduced in (2.5). Examples more familiar to combinatorialists will be considered in chapter IV.

For sheaflike categorical structures the section coefficients $G(t;t_1,t_2)$ and the derived numbers $G(t;1)$ from chapter I, §1, have special properties.

(3.12) THEOREM: For any types $v,w,t_1,t_2 \in T$,

$$G(v+w;t_1,t_2) = \Sigma\, G(v;v_1,v_2)G(w;w_1,w_2)$$

where the summation is taken over all types $v_1,v_2,w_1,w_2 \in T$ such that $v_1+w_1=t_1$ and $v_2+w_2=t_2$ (By (3.7) this sum is finite).
Thus the numbers $G(t;t_1,t_2)$ are "bisection coefficients" in the sense of G.C.Rota ([JR,79],p.97).

(3.13) REMARK: In Example (1.53) dealing with subsets of finite sets, the section coefficients are the binomial coefficients and formula (3.12) reduces to the Vandermonde convolution identity ([Co,74]).
In most examples however, there is no simple expression for the section coefficients $G(t;t_1,t_2)$ unless t is an indecomposable type. Then formula (3.12) can be used recursively to calculate the numbers $G(t;t_1,t_2)$, $t \notin U$, from the <u>elementary section coefficients</u> $G(u;t_1,t_2)$, $u \in U$.

Proof of Theorem (3.12): When $v=0$ or $w=0$, the formula is evident as $G(0;v_1,v_2)=1$ if $v_1=0=v_2$ and $G(0;v_1,v_2)=0$ else. Hence suppose that both v and w are non-zero. Choosing representatives $q:X_1 \to Y_1$, $r:X_2 \to Y_2$ of v or w resp., we get the representative $s=q \oplus r:X_1 \oplus X_2 \to Y_1 \oplus Y_2$ of $v+w$. In the first part of the proof it is shown that the map

$$\begin{array}{ccc} \{\langle q_1,q_2\rangle \in S_2\langle M\rangle ; q_2q_1=q\} \times \{\langle r_1,r_1\rangle \in S_2\langle M\rangle ; r_2r_1=r\} & & (\langle q_1,q_2\rangle,\langle r_1,r_2\rangle) \\ \downarrow B & & \downarrow \\ \{\langle s_1,s_2\rangle \in S_2\langle M\rangle ; s_2s_1=s\} & & (\langle q_1\oplus r_1,q_2\oplus r_2\rangle) \end{array}$$

is well-defined and a bijection. Then, in the second part, a simple counting argument yields the result.

(i) B is well-defined: If the diagrams

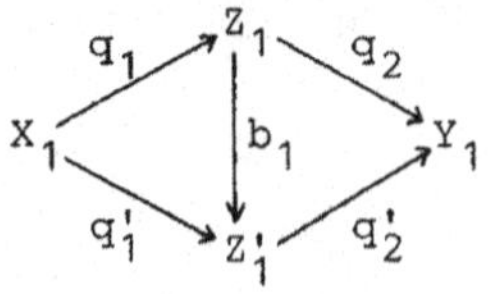

and

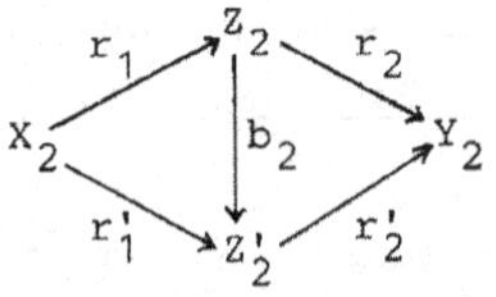

commute, then also

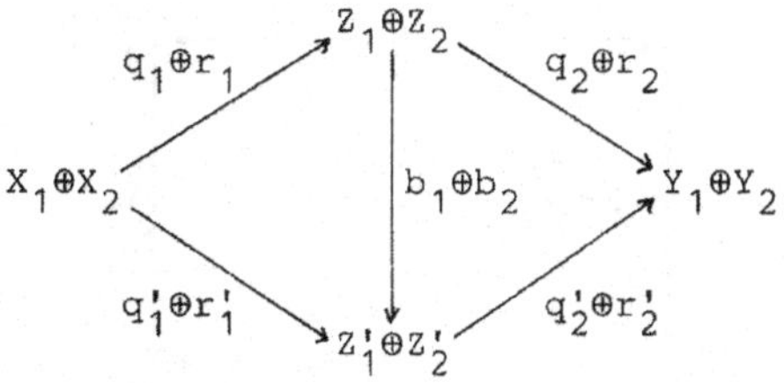

commutes. Thus $<q_1,q_2>=<q_1',q_2'>$ and $<r_1,r_2>=<r_1',r_2'>$ imply that $<q_1\oplus r_1,q_2\oplus r_2>=<q_1'\oplus r_1',q_2'\oplus r_2'>$.

B is one-to-one: Suppose that $<q_1\oplus r_1,q_2\oplus r_2>=<q_1'\oplus r_1',q_2'\oplus r_2'>$. Then there is an isomorphism b such that the diagram

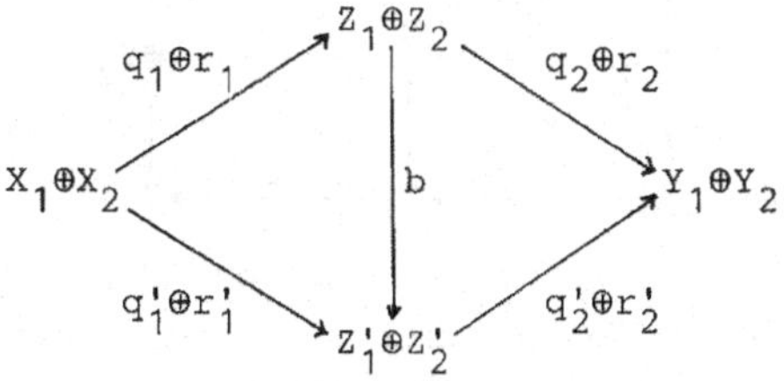

(*)

commutes. We apply (M6) to the morphism $q_2\oplus r_2$ and to the partition $\{[Y_1\to Y_1\oplus Y_2],[Y_2\to Y_1\oplus Y_2]\}$ of Y. By the uniqueness assertion, it follows from the diagrams

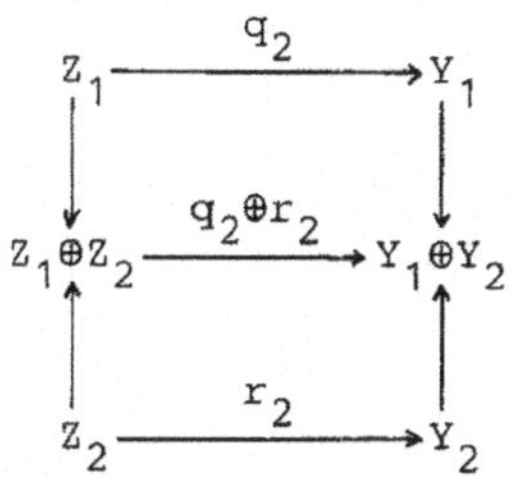

and

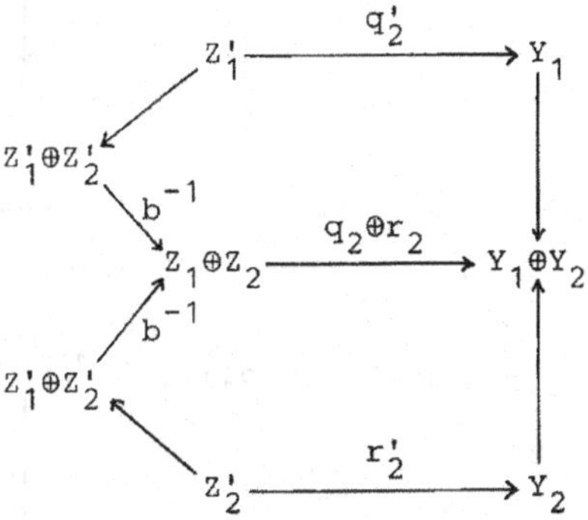

that $[q_2\to q_2\oplus r_2]=[q_2'\to q_2\oplus r_2]$ and $[r_2\to q_2\oplus r_2]=[r_2'\to q_2\oplus r_2]$ in $\mathrm{Sub}(q_2\oplus r_2)$. Hence there exist isomorphisms $b_1:Z_1\to Z_1'$ and $b_2:Z_2\to Z_2'$ such that $b=b_1\oplus b_2$. Looking back at diagram (*), we see that both

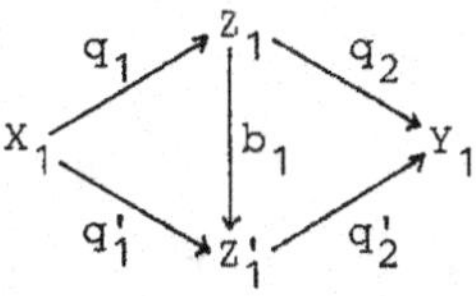

and

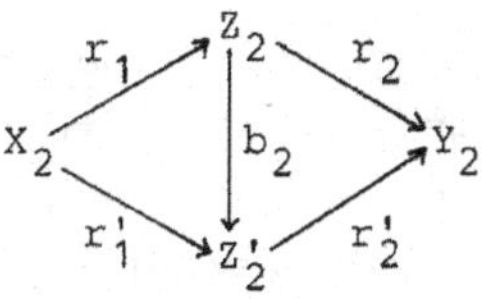

commute. Thus $<q_1,q_2>=<q_1',q_2'>$ and $<r_1,r_2>=<r_1',r_2'>$.

B is onto: Given a simplex $(s_1,s_2)\in S_2(M)$ such that $s_2s_1=s$,

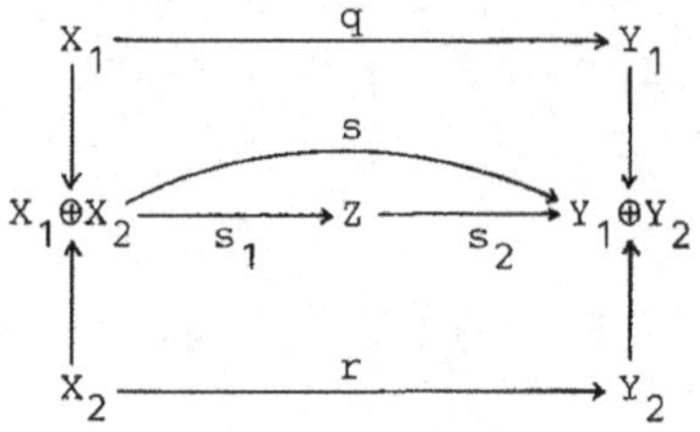

we get, applying (M6) twice, morphisms q_2, r_2 in M such that $s_2 \simeq q_2 \oplus r_2$, and morphisms q_1, r_1 in M such that $s_1 \simeq q_1 \oplus r_1$.

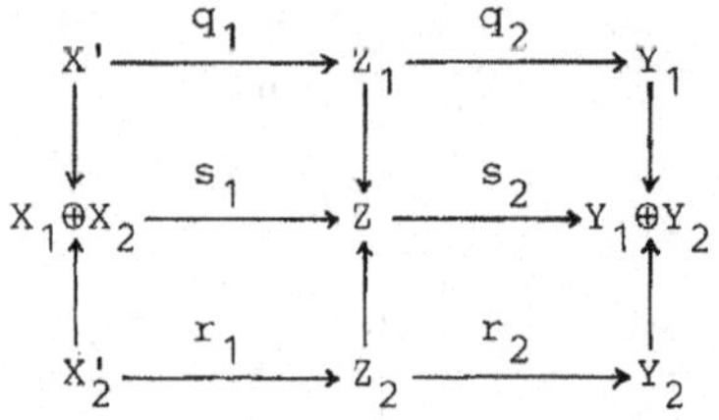

Because $q \oplus r = s = s_2 s_1 \simeq q_2 q_1 \oplus r_2 r_1$, we have, by (M6), $[q \to s] = [q_2 q_1 \to s]$ and $[r \to s] = [r_2 r_1 \to s]$ in Sub(s). Consequently, q_1 and r_1 can be chosen in such a way that $q = q_2 q_1$, $r = r_2 r_1$ and that the diagram

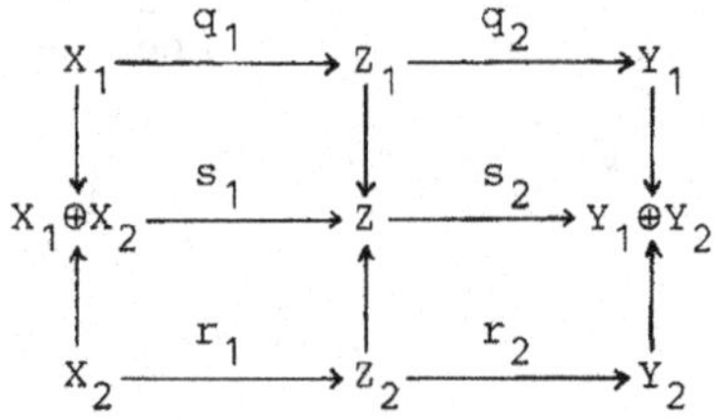

commutes. Then the induced diagram

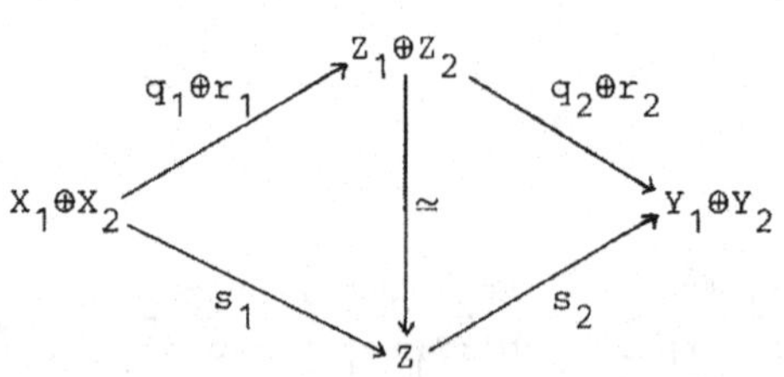

shows that $<s_1, s_2> = <q_1 \oplus r_1, q_2 \oplus r_2> = B(<q_1, q_2>, <r_1, r_2>)$.

(ii) Under the bijection B, the subset

$\{<s_1, s_2> \in S_2<M>; s_2 s_1 = s, \bar{s}_1 = t_1, \bar{s}_2 = t_2\}$

is the image of the disjoint union of the sets

$\{<q_1, q_2> \in S_2<M>; q_2 q_1 = q, \bar{q}_1 = v_1, \bar{q}_2 = v_2\} \times \{<r_1, r_2> \in S_2<M>; r_2 r_1 = r, \bar{r}_1 = w_1, \bar{r}_2 = w_2\}$

indexed by all $(v_1, v_2), (w_1, w_2) \in T \times T$ such that $v_1 + w_1 = t_1$ and $v_2 + w_2 = t_2$. Counting on both sides, we obtain the stated formula. □

For the higher section coefficients $G(t;t_1,..,t_l)$ an analogous formula can be derived.

(3.14) COROLLARY: Let $l \geq 3$. Then, for any types $v,w,t_1,..,t_l \in T$,

$$G(v+w;t_1,..,t_l) = \Sigma\, G(v;v_1,..,v_l)G(w;w_1,..,w_l)$$

where the sum ranges over all $(v_1,..,v_l),(w_1,..,w_l) \in T^l$ such that $v_1+w_1=t_1,..,v_l+w_l=t_l$.

Proof: We use induction on l. By the recursion formula (1.9) and Theorem (3.12) we have $G(v+w;t_1,..,t_l) = \Sigma_x G(x;t_1,..,t_{l-1})G(v+w;x,t_l) =$
$= \Sigma_x G(x;t_1,..,t_{l-1})\Sigma_{y+z=x,v_l+w_l=t_l} G(v;y,v_l)G(w;z,w_l) =$
$= \Sigma_{y,z,v_l+w_l=t_l} G(y+z;t_1,..,t_{l-1})G(v;y,v_l)G(w;z,w_l) =$
$= \Sigma_{y,z}\Sigma_{v_i+w_i=t_i} G(y;v_1,..,v_{l-1})G(z;w_1,..,w_{l-1})G(v;y,v_l)G(w;z,w_l) =$
$= \Sigma_{v_i+w_i=t_i} G(v;v_1,..,v_l)G(w;w_1,..,w_l)$. □

In (1.11) we have introduced, for a type t and an integer $l \geq 0$, the number $G(t;l)$ which counts the factorizations of length l of any representative of t up to simplicial equivalence. Furthermore, the dimension of a type t has been characterized as the largest integer $l \geq 0$ such that $G(t;l) \neq 0$.

(3.15) PROPOSITION: For any types $v,w \in T$ and any integer $l \geq 0$,

$$G(v+w;l) = \Sigma \binom{l}{l-m,\,l-n,\,m+n-l} G(v;m)G(w;n)$$

where the summation is taken over all integers $m,n \geq 0$ such that $m,n \leq l \leq m+n$. Here $\binom{l}{l-m,\,l-n,\,m+n-l} = \frac{l!}{(l-m)!(l-n)!(m+n-l)!}$ is the trinomial coefficient.

Proof: This formula could be proved here but is a special case of Lemma (3.24) in the next section. □

(3.16) COROLLARY: For any types $v,w \in T$, $\dim(v+w) = \dim(v)+\dim(w)$. Thus the dimension function $\dim: T \to \mathbb{N}_0$, $t \to \dim(t)$, is a homomorphism of monoids. □

§2. The incidence bialgebra and the affine monoid of multiplicative functions

Let $\underline{K}, M, \sim$ be a sheaflike categorical structure with set of types T and section coefficients $G(t;t_1,t_2)$. Let k be a commutative ring, and let $H = k(T;G)$ denote the incidence algebra of $\underline{K}, M, \sim$. In the sequel we use the basic theory of bialgebras and affine monoids sketched in the appendix, §3.

Since T is a commutative monoid, $H = k^T$ becomes a cocommutative topological coalgebra with the comultiplication

$$\Delta : H \to H\hat{\otimes}H \ , \ e(t) \to \Sigma \ e(t_1)\hat{\otimes}e(t_2)$$

where the sum runs over all types $t_1, t_2 \in T$ such that $t_1 + t_2 = t$, and the counit

$$\varepsilon : H \to k \ , \ e(t) \to \delta_{t,0} \text{ (the Kronecker delta) }.$$

The following theorem shows that the algebra and the coalgebra structure on k^T are compatible.

(3.17) THEOREM: The maps Δ and ε are k-algebra homomorphisms and turn the incidence algebra $H = k(T;G)$ into a cocommutative topological bialgebra which we call the <u>incidence bialgebra</u> of $\underline{K}, M, \sim$.

Proof: The unit of H is the delta function $\delta = \Sigma_{t\in T'} e(t)$, so $\varepsilon(\delta) = 1$ and $\Delta(\delta) = \Sigma_{t\in T'} \Sigma_{t_1+t_2=t} e(t_1)\hat{\otimes}e(t_2) = \Sigma_{t_1,t_2\in T'} e(t_1)\hat{\otimes}e(t_2) = \delta\hat{\otimes}\delta$ by (3.8). To check that Δ and ε preserve the convolution product, let $v, w \in T$. Then $e(v)e(w) = \Sigma_t G(t;v,w)e(t)$,

$\varepsilon(e(v)e(w)) = G(0;v,w) = \delta_{v,0}\delta_{w,0} = \varepsilon(e(v))\varepsilon(e(w))$ and

$\Delta(e(v))\Delta(e(w)) = (\Sigma_{v_1+v_2=v} e(v_1)\hat{\otimes}e(v_2))(\Sigma_{w_1+w_2=w} e(w_1)\hat{\otimes}e(w_2)) =$

$= \Sigma_{v_1+v_2=v, w_1+w_2=w} e(v_1)e(w_1)\hat{\otimes}e(v_2)e(w_2)$.

As $e(v_i)e(w_i) = \Sigma_{t_i} G(t_i;v_i,w_i)e(t_i)$ for $i=1,2$, it follows from (3.12) that $\Delta(e(v))\Delta(e(w)) =$

$= \Sigma_{t_1,t_2}(\Sigma_{v_1+v_2=v, w_1+w_2=w} G(t_1;v_1,w_1)G(t_2;v_2,w_2))e(t_1)\hat{\otimes}e(t_2) =$

$= \Sigma_{t_1,t_2} G(t_1+t_2;v,w)e(t_1)\hat{\otimes}e(t_2) = \Sigma_t G(t;v,w)\Sigma_{t_1+t_2=t} e(t_1)\hat{\otimes}e(t_2) =$

$= \Sigma_t G(t;v,w)\Delta(e(t)) = \Delta(e(v)e(w))$. □

(3.18) COROLLARY: Suppose that the ground ring k has characteristic 0. Then the topological bialgebra H is graded by the sequence of closed submodules

$$V(d) = \{f \in k^T ; \ f(t) = 0 \text{ unless } \dim(t) = d\} \quad , \quad d = 0,1,2,.. \quad ,$$

if and only if, for every morphism s in M, all maximal factorizations of s have equal length.

Proof: By Corollary (3.16) $\Delta V(d) \subset \Sigma_{i=0}^{d} V(i)\hat{\otimes}V(d-i)$ for all d, and $\varepsilon(V(d))=0$ unless $d=0$, so Theorem (1.20) yields the conclusion. □

Now recall that H is a topological k-module with the topological basis $(e(t);t\in T)$. Let $A=H'$ be the dual k-module with the dual basis $(x(t);t\in T)$. Then, by the duality

(3.19) $\qquad H\times A \to k \ , \ (e(v),x(w)) \to \delta_{v,w}$,

A becomes a commutative abstract bialgebra with

the multiplication $\quad x(t_1)x(t_2) = x(t_1+t_2)$,

the unit $\quad x(0)$,

the comultiplication $\quad \Delta: A \to A\otimes A$,

$$\Delta(x(t)) = \Sigma_{t_1,t_2\in T} G(t;t_1,t_2)x(t_1)\otimes x(t_2) \ ,$$

and the counit $\quad \varepsilon: A \to k$,

$\varepsilon(x(t))=1$ if $t\in T'$ and $\varepsilon(x(t))=0$ otherwise .

In particular, as an algebra, A is the monoid algebra of T, i.e. the polynomial algebra in the indeterminates $x(u)$, $u\in U$ (compare [JR,79],pp.97).

Furthermore, if H is graded by the sequence $(V(d))_{d\geq 0}$, then the bialgebra A is graded by the sequence $(W(d))_{d\geq 0}$ where $W(d)$ is the k-submodule of A generated by the $x(t)$ with $\dim(t)=d$.

(3.20) DEFINITION: Let R be a commutative k-algebra with unit 1. A function $f:T\to R$ is called <u>multiplicative</u> if

$$f(0)=1 \qquad \text{and} \qquad f(v+w) = f(v)f(w) \quad \text{for all } v,w\in T \ .$$

Then f is given by its values on the indecomposable types:

$$f(t) = \Pi_{u\in U} f(u)^{t(u)} \qquad \text{for} \qquad t = \Sigma_{u\in U} t(u)u \ .$$

Let Mu(R) denote the set of multiplicative functions on T with values in R.

The preceding considerations immediately imply

(3.21) PROPOSITION: For every commutative k-algebra R, Mu(R) is a submonoid of the incidence algebra R(T;G) with respect to the convolution multiplication. The monoid functor

$$\mathrm{Mu}: \underline{\mathrm{Al}}_k \to \underline{\mathrm{Mo}} \ , \ R \to \mathrm{Mu}(R) \ ,$$

is represented by the polynomial algebra $A= k[x(u);u\in U]$ through

$$\mathrm{Al}_k(A,R) \to \mathrm{Mu}(R) \ , \ g \to f, \ f(t)=g(x(t)) \ ,$$

and hence is k-free. We call Mu the <u>affine monoid of multiplicative functions</u> and identify its covariant bialgebra A* with the incidence bialgebra H by the duality (3.19). □

Summing up, we have derived three equivalent algebraic structures from the sheaflike categorical structure $\underline{K}, M, \sim$:

the cocommutative topological bialgebra H ,

the commutative discrete bialgebra A ,

and the affine monoid Mu .

For combinatorial purposes the affine monoid Mu seems to be best suited, because often a power series representation of Mu exists which can be used to set up generating functions. Moreover, as

$$(3.22) \qquad fg(u) = \Sigma_{t_1,t_2 \in T} G(u;t_1,t_2) f(t_1) g(t_2) \quad , \quad u \in U \quad ,$$

the product of two multiplicative functions can be computed if one only knows the elementary section coefficients $G(u;t_1,t_2)$, $u \in U$.

(3.23) PROPOSITION: Let R be a commutative k-algebra with unit 1. Then

(i) an $f \in Mu(R)$ is invertible in $R(T;G)$ if and only if $f(u)$ is invertible in R for every $u \in U'$. In this case f^{-1} also is multiplicative.

(ii) Mu(R) contains the submonoid

$$Mu'(R) = \{f \in Mu(R);\ f(u)=0 \text{ for all } u \in U''\}$$

which is isomorphic to the product monoid $R^{U'}$ by $f \to (f(u))_{u \in U'}$, and the subgroup

$$Mu''(R) = \{f \in Mu(R);\ f(u)=1 \text{ for all } u \in U'\} \quad .$$

(iii) E(Mu(R)), the group of invertible elements of Mu(R), is the semidirect product of the subgroup E(Mu'(R)) and the normal subgroup Mu"(R). The corresponding decomposition of an $f \in E(Mu(R))$ is $f = gh$ where $g = \Sigma_{v \in T'} f(v)e(v) \in E(Mu(R))$ and

$$h = \delta + \Sigma_{w \in T(1)} \frac{f(w)}{f(\mathrm{dom}(w))} e(w) \in Mu''(R) \quad .$$

The operation of E(Mu'(R)) on Mu"(R) by inner automorphisms is given by $ghg^{-1}(t) = \frac{g(\mathrm{dom}(t))}{g(\mathrm{cod}(t))} h(t) \quad , \quad t \in T \quad .$

Proof: The point is that Mu"(R) is closed under inversion which will be shown at the end. Then, recalling Corollary (1.21) and Proposition (1.22), (ii) is evident. For any multiplicative function $f \in E(R(T;G))$, the component of f in $E(V \hat{\otimes} R)$, $g = \Sigma_{v \in T'} f(v)e(v)$, is multiplicative and thus contained in E(Mu'(R)). This establishes (iii). To prove (i), let $f = gh$ be the decomposition of f according to $E(Mu(R)) = E(Mu'(R))Mu''(R)$. Then $f^{-1} = h^{-1}g^{-1}$ is multiplicative because so are $g^{-1} = \Sigma_{v \in T'} f(v)^{-1} e(v)$ and h^{-1}. It remains to demonstrate that $f^{-1} \in G''(R)$ for every $f \in G''(R)$. Applying (A.4) to $p = f - \delta \in I(1) \hat{\otimes} R$, we get $f^{-1} = \Sigma_{l=0}^{\infty} (-1)^l p^l$.

Now let $v, w \in T$. By the following Lemma,

$$f^{-1}(v+w) = \Sigma_{m,n\geq 0} \ \Sigma_{m,n\leq l\leq m+n} (-1)^{l} \binom{l}{l-m,l-n,m+n-l} \ p^{m}(v)p^{n}(w) \ .$$

But the inner sum is $(-1)^{m+n}$ as can be seen by developing $(1+X+Y+XY)^{-1}$ and $(1+X)^{-1}(1+Y)^{-1}$ into geometric series and comparing coefficients at X^mY^n. Consequently,

$$f^{-1}(v+w) = (\Sigma_{m=0}^{\infty}(-1)^{m}p^{m}(v))(\Sigma_{n=0}^{\infty}(-1)^{n}p^{n}(w)) = f^{-1}(v)f^{-1}(w)$$

which completes the proof. □

(3.24) LEMMA: Let $f\in G''(R)$ and $p= f-\delta$. Then, for any $l\geq 0$ and $v,w\in T$,

$$p^{l}(v+w) = \Sigma \binom{l}{l-m,l-n,m+n-l} p^{m}(v)p^{n}(w)$$

where the summation is taken over all integers $m,n\geq 0$ such that $m,n\leq l\leq m+n$.

(In particular, if $f=\zeta$ is the zeta function and $p= \zeta-\delta = \eta$, we recover Proposition (3.15) because $\eta^{l}(t)= G(t;l)$ by (1.24).)

Proof: First observe that $p(t)=0$ if $t\in T'$ and $p(t)=f(t)$ if $t\in T(1)$. Since the formula is obvious for $l=0$ or $l=1$, we can assume that $l\geq 2$. Using the vector notation $\underline{t}=(t_1,..,t_l)\in T^{l}$, we have

$$p^{l}(v+w)= \Sigma_{\underline{t}\in T(1)^l} G(v+w;\underline{t})f(t_1)..f(t_l)=$$

$$= \Sigma_{\underline{v}+\underline{w}\in T(1)^l} G(v;\underline{v})G(w;\underline{w})f(v_1)..f(v_l)f(w_1)..f(w_l)$$

by (1.15) and (3.14). With the abbreviations $S(\underline{v})=\{i;v_i\in T(1)\}$ and $S(\underline{w})=\{j;w_j\in T(1)\}$ it follows that

$$p^{l}(v+w) =$$

$$= \Sigma_{B,C} (\Sigma_{\underline{v},S(\underline{v})=B}G(v;\underline{v})f(v_1)..f(v_l)) \ (\Sigma_{\underline{w},S(\underline{w})=C}G(w;\underline{w})f(w_1)..f(w_l))$$

where the outer sum runs over all subsets B,C of $\{1,..,l\}$ whose union is $\{1,..,l\}$. Substituting $G(v;\underline{v})= e(v_1)..e(v_l)(v)$ into the second sum, we obtain $\Sigma_{\underline{v}\in T^l,S(\underline{v})=B}f(v_1)e(v_1)..f(v_l)e(v_l)(v)= p^{\#B}(v)$ because $\Sigma_{x\in T'}f(x)e(x)=\delta$ and $\Sigma_{x\in T(1)}f(x)e(x)=p$.

Similarly, the third sum is $p^{\#C}(w)$. Finally, given integers $m,n\geq 0$ with $m,n\leq l\leq m+n$, the number of pairs (B,C) of subsets of $\{1,..,l\}$ such that $B\cup C=\{1,..,l\}$, $\#B=m$ and $\#C=n$, is just $\binom{l}{l-m,l-n,m+n-l}$ which implies the stated formula. □

The <u>multiplicative monoid</u> is the affine monoid

$$\underline{Al}_k \to \underline{Mo} \ , \ R \to R \text{ with multiplication} \ .$$

Combining (3.23) with Theorem (1.23), we get the following result.

(3.25) THEOREM: Suppose that k is a field.
The affine monoid of multiplicative functions Mu contains the closed submonoid

$$Mu': \underline{Al}_k \to \underline{Mo}\ ,\ R \to Mu'(R)\ ,$$

which is isomorphic to the U'-fold product of the multiplicative monoid, and the closed unipotent subgroup

$$Mu": \underline{Al}_k \to \underline{Gr}\ ,\ R \to Mu"(R)\ .$$

The affine group of invertible multiplicative functions E(Mu) is the semidirect product of the closed diagonalizable subgroup E(Mu') and the closed unipotent normal subgroup Mu", and hence is triangulable. □

So the algebraically interesting part of Mu is the unipotent group Mu". Since $Mu" = Mu \cap \Lambda$ where the group Λ has been defined in (1.23), the structure of Λ described in (A.22) is inherited by Mu".

(3.26) PROPOSITION: Suppose that k is a field.
The dimension of types induces the filtration of Mu"

$$Mu"=Mu"(1) \supset Mu"(2) \supset Mu"(3) \supset \ldots$$

by the closed affine normal subgroups

$$Mu"(d)(R) = \{f \in Mu"(R);\ f(u)=0 \text{ for all } u \in U(1,d)\}$$

where $U(1,d) = \{u \in U;\ 1 \leq \dim(u) \leq d-1\}$. Moreover, Mu" is the inverse limit of the unipotent affine groups Mu"/Mu"(d) whose underlying schemes are isomorphic to the affine spaces $\mathbb{A}^{U(1,d)}$ by

$$Mu"(R)/Mu"(d)(R) \to \mathbb{A}^{U(1,d)}(R)=R^{U(1,d)}\ ,\ f \to (f(u))_{u \in U(1,d)}\ .$$

(In particular, if the sets U(1,d) are finite, then the groups Mu"/Mu"(d) are algebraic.) □

We next determine the Lie algebra of Mu".
From (A.25) we know that the Lie algebra of Λ is H(1), so Lie(Mu") is a sub Lie algebra of H(1). An element $h \in H(1)$ is in Lie(Mu") if and only if $\delta+\tau h: T \to k[\tau]$ is multiplicative ($k[\tau]$= the algebra of dual numbers) which is equivalent to the following condition:
For any types $t_1,t_2 \in T$, $h(t_1+t_2) = \delta(t_1)h(t_2)+\delta(t_2)h(t_1)$.
But this says that $h(t)=0$ unless $t=t'+u$ for some $t' \in T'$ and some $u \in U"$, and that $h(t'+u)=h(u)$ for all $t' \in T'$ and all $u \in U"$. Then

$$h= \Sigma_{t \in T(1)} h(t)e(t) = \Sigma_{t' \in T', u \in U"} h(t'+u)e(t'+u) = \Sigma_{u \in U"} h(u)e"(u)$$

where $e"(u) = \Sigma_{t' \in T'} e(t'+u)$. Thus Lie(Mu") has the topological k-basis $(e"(u); u \in U")$. The coefficient of the bracket $[e"(u_1),e"(u_2)]$ at e"(u) is $e"(u_1)e"(u_2)(u)-e"(u_2)e"(u_1)(u) =$

$$= \Sigma_{t_1',t_2' \in T'} \{G(u;t_1'+u_1,t_2'+u_2)-G(u;t_2'+u_2,t_1'+u_1)\}\ .$$

But $G(u;t_1'+u_1,t_2'+u_2)=0$ unless $dom(u)=t_1'+dom(u_1)$ and $cod(u)=t_2'+cod(u_2)$. This suggests a definition.

(3.27) DEFINITION: For $u,u_1,u_2 \in U''$, set

$$G''(u;u_1,u_2) = G(u;u_1+dom(u)-dom(u_1),u_2+cod(u)-cod(u_2))$$

if $dom(u) \geq dom(u_1)$ and $cod(u) \geq cod(u_2)$ in the componentwise order on $T' = \oplus_{u' \in U'} \mathbb{N}_0 u'$, and

$$G''(u;u_1,u_2)=0 \quad \text{else} .$$

Hence, by the calculation above,
$[e''(u_1),e''(u_2)](u) = G''(u;u_1,u_2)-G''(u;u_2,u_1)$.
Combining these results with Corollary (A.25), we have proved

(3.28) PROPOSITION: The Lie algebra of Mu", Lie(Mu")⊂H, is a topologically nilpotent Lie algebra. With respect to the topological k-basis

$$e''(u) = \Sigma_{t' \in T'} e(t'+u) \quad , \quad u \in U'' \quad ,$$

the Lie bracket is given by

$$[e''(u_1),e''(u_2)] = \Sigma_{u \in U''} \{G''(u;u_1,u_2)-G''(u;u_2,u_1)\} e''(u) \quad .$$

If k contains the rational numbers, then Mu" is isomorphic to the affine group

$$\underline{Al}_k \to \underline{Gr} \ , \ R \to Lie(Mu'')\hat{\otimes}_k R \ ,$$

where the group structure is the Campbell-Hausdorff composition. Therefore, if $k \supset \mathbb{Q}$, the affine group Mu" is determined up to isomorphism by the numbers

$$G''(u;u_1,u_2)-G''(u;u_2,u_1) \quad , \quad u,u_1,u_2 \in U'' \quad . \quad \square$$

(3.29) REMARK: Consider the special situation that all isomorphisms in $\underline{K}$ have the same type. Then $T'=\{0\}$, U' is empty and U=U" . Here the affine monoid of multiplicative functions Mu=Mu" is a group, the incidence bialgebra H=k(T;G) is a topological Hopf algebra and the Lie algebra Lie(Mu)=Lie(Mu") has the topological basis $(e(u);u \in U)$ and the Lie bracket

$$[e(u_1),e(u_2)] = \Sigma_{u \in U} \{G(u;u_1,u_2)-G(u;u_2,u_1)\} e(u) \quad .$$

The antipode of H S:H→H is dual to the antipode of A S:A→A under the duality (3.19). Inverting the universal element
$\delta + \Sigma_{t \neq 0} x(t)e(t) \in Mu(A)$ by (A.14) yields the formula

$$S(x(u)) = \Sigma_{t \neq 0} a(u;t) x(t) \quad , \quad u \in U \quad ,$$

where $a(u;t) = \Sigma (-1)^{|\gamma|} G(u;\gamma)$. Here the sum runs over all $\gamma \in \mathbb{N}_0(T(1))$ such that $\Sigma_w \gamma(w)w=t$, and $G(u;\gamma)$ is defined in (A.12).
Finally, assume that for every morphism s in M all maximal factorizations of s have equal length. Then $a(u;t)=0$ unless $dim(u)=dim(t)$

because the $G(u;\gamma)$ count factorizations. Thus S leaves the k-submodule $W(d)$ of A generated by the $x(d)$ with $\dim(t)=d$ stable, for every d. Dually, $S(V(d))=V(d)$ for all d which shows that the topological Hopf algebra H is graded by the sequence $(V(d))_{d\geq 0}$.

(3.30) Zeta, Möbius and characteristic function:
In (1.24) we have introduced the combinatorically important functions $\zeta,\mu,D^{\dim}$ and χ. Obviously the zeta function $\zeta:T\to k$, $\zeta(t)=1$ for all t, is multiplicative and contained in $Mu''(k)$. Hence its inverse, the Möbius function $\mu=\zeta^{-1}$ is multiplicative, too, and completely determined by the values $\mu(u)$, $u\in U''$. Furthermore, by (3.16), also $D^{\dim}\in Mu''(k[D])$ which implies that both the characteristic function $\chi=\mu D^{\dim}$ and the function $D^{\dim}\zeta$ lie in $Mu''(k[D])$.

(3.31) REMARK: Let $\underline{K},M,\sim$ be a categorical structure. In (1.28) we have seen that the incidence algebra $k(T;G)$ of $\underline{K},M,\sim$ is isomorphic to a closed subalgebra of the incidence algebra $k(I;\Gamma)$ of the categorical structure $\underline{K},M,\simeq$ where $\simeq$ denotes the isomorphism relation on M. If $\underline{K},M,\sim$ and thus $\underline{K},M,\simeq$ are sheaflike, then can: $I\to T$, $\hat{s}\to\bar{s}$, is a homomorphism of monoids and induces an isomorphism from the affine monoid of multiplicative functions on T onto a closed submonoid of the affine monoid of multiplicative functions on I.

(3.32) Products of sheaflike categorical structures: Let $\underline{K}_1,M_1,\sim_1$ and $\underline{K}_2,M_2,\sim_2$ be two sheaflike categorical structures. From (1.29) we know that the incidence algebra $k(T;G)$ of the categorical structure

$$\underline{K}=\underline{K}_1\times\underline{K}_2\quad,\quad M=M_1\times M_2\quad,\quad \sim\;=\;\sim_1\times\sim_2$$

is isomorphic to the topological tensor product of the incidence algebras $k(T_1;G_1)$ and $k(T_2;G_2)$. Moreover $\underline{K},M,\sim$ is sheaflike, the monoid of types T is isomorphic to the product of the monoids T_1 and T_2, and the affine monoid of multiplicative functions on T is isomorphic to the product of the affine monoids of multiplicative functions on T_1 and T_2.

(3.33) Substructures: Let $\underline{K},M,\sim$ be a sheaflike categorical structure, and let M_o be a subclass of M which is saturated with respect to $\sim$ and has the properties (M1) and (M4). Then, by (1.30), the incidence algebra $k(T_o;G_o)$ of the categorical structure $\underline{K},M_o,\sim$ is isomorphic to a closed subalgebra of $k(T;G)$. Furthermore $\underline{K},M_o,\sim$ is sheaflike, the set T_o of types of morphisms in M_o is a submonoid of T, and the affine monoid of multiplicative functions on T_o is isomorphic to a closed submonoid of Mu.

§3. Subobjects and quotient objects

Let $\underline{K}, M, \sim$ be a sheaflike categorical structure where M is a class of monomorphisms with (S2) or a class of epimorphisms with (Q2). In this section we consider the partially ordered sets of subobjects or quotient objects in M. Whereas in chapter I, §3, the two cases were dual to each other, this duality no longer exists for the axioms of a sheaflike categorical structure are not self-dual.
Let $\underset{\sim}{P}$ be a system of representatives of the isomorphism classes of indecomposable objects in $\underline{K}$.

First case: M consists of monomorphisms.

(3.34) PROPOSITION: Let Y be an object in $\underline{K}$ of Krull-Schmidt type $m \in \mathbb{N}_0(\underset{\sim}{P})$. Then the poset $\mathrm{Sub}_M(Y)$ is order-isomorphic to the product

$$\Pi_{P \in \underset{\sim}{P}} \mathrm{Sub}_M(P)^{m(P)} \quad .$$

Proof: This follows inductively from the Lemma below. □

(3.35) LEMMA: For any objects Y_1, Y_2 in $\underline{K}$, the map

$$\Psi\colon \mathrm{Sub}_M(Y_1) \times \mathrm{Sub}_M(Y_2) \to \mathrm{Sub}_M(Y_1 \oplus Y_2) \quad , \quad ([s_1],[s_2]) \to [s_1 \oplus s_2] \quad ,$$

is an order isomorphism.

Proof: Ψ clearly is well-defined, onto by (M6) and order-preserving. It remains to show that if $[r_1 \oplus r_2] \le [s_1 \oplus s_2]$ in $\mathrm{Sub}_M(Y_1 \oplus Y_2)$, then $[r_1] \le [s_1]$ in $\mathrm{Sub}_M(Y_1)$ and $[r_2] \le [s_2]$ in $\mathrm{Sub}_M(Y_2)$. Hence suppose that there is a monomorphism q in M which makes the diagram

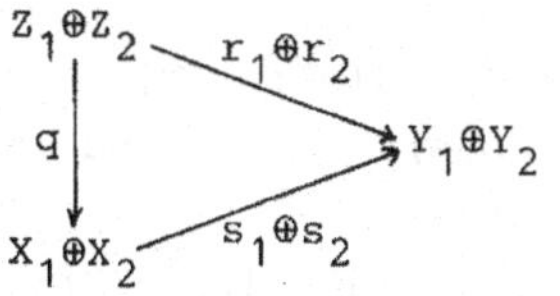

commute. Applying (M6) to the morphism q and to the partition $\{[X_i \to X_1 \oplus X_2]; i=1,2\}$ of $X_1 \oplus X_2$, we get a commuting diagram

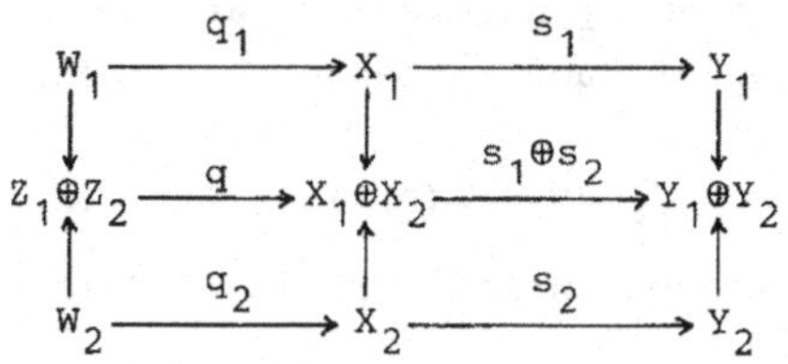

.

Then $\{[s_iq_i \to r_1 \oplus r_2]; i=1,2\}$ and $\{[r_i \to r_1 \oplus r_2]; i=1,2\}$ are two partitions of $[r_1 \oplus r_2]$ inducing the same image partition of $Y_1 \oplus Y_2$. By (M6) we conclude that $[s_iq_i \to r_1 \oplus r_2] = [r_i \to r_1 \oplus r_2]$ in $\mathrm{Sub}(r_1 \oplus r_2)$ and that $[r_i] \leq [s_i]$ in $\mathrm{Sub}_M(Y_i)$ for $i=1,2$. □

An immediate consequence of (1.36) and (3.18) is

(3.36) COROLLARY: Suppose that the ground ring k has characteristic 0. Then the topological bialgebra H is graded by the sequence $(V(d))_{d \geq 0}$ if and only if the posets $\mathrm{Sub}_M(Y)$, Y an object in $\underline{K}$, satisfy the (Jordan-Dedekind) chain condition. □

<u>Second case:</u> M consists of epimorphisms.

Here we have no analogue of Proposition (3.34), but we will set up a generating function for the number of quotient objects. Before we state a result which follows at once from (1.43) and (3.18).

(3.37) COROLLARY: Suppose that the ground ring k has characteristic 0. Then the topological bialgebra H is graded by the sequence $(V(d))_{d \geq 0}$ if and only if the posets $\mathrm{Qu}_M(X)$, X an object in $\underline{K}$, satisfy the (Jordan-Dedekind) chain condition. □

Finally, we turn to the problem of counting quotient objects in M. In the sequel we suppose that $\underline{K}$ has finite automorphism groups and that all posets $\mathrm{Qu}_M(X)$ are finite.
For any Krull-Schmidt type $n \in \mathbb{N}_0(\underset{\sim}{P})$ we put $X(n) = \oplus_P n(P)P$.
If $m \in \mathbb{N}_0(\underset{\sim}{P})$ is another KS-type, then the automorphism group of $X(m)$ acts freely on

$$M(n,m) = \{s: X(n) \to X(m);\ s \in M\}$$

by $\mathrm{Aut}(X(m)) \times M(n,m) \to M(n,m)$, $(b,s) \to bs$, and induces a bijection between orbits and quotient objects

$$\begin{aligned} \mathrm{Aut}(X(m)) \backslash M(n,m) &\to \{[r: X(n) \to Y] \in \mathrm{Qu}_M(X);\ \text{KS-type of } Y = m\} \\ \mathrm{Aut}(X(m))s &\to [s] \end{aligned} \tag{3.38}$$

Therefore, $\#\mathrm{Qu}_M(X(n)) = \Sigma_m \#M(n,m)/a(m)$, and all sets $M(n,m)$ are finite. Of course, $M(0,m)$ is empty unless $m=0$, and $M(0,0) = \{0 \to 0\}$ has one element. Besides, the number of quotient objects $[r: X(n) \to Y]$ of $X(n)$ where Y is indecomposable is

$$iq(n) = \Sigma_{P \in \underset{\sim}{P}} \#M(n, \varepsilon(P))/a(\varepsilon(P)) .$$

As a new application of the exponential formula (2.25) we have the following identity.

(3.39) PROPOSITION: In the power series algebra $\mathbb{Q}[[z(P),w(P);P\in\underset{\sim}{P}]]$,

$$\Sigma_{n,m} \#M(n,m)\ z^n/n!\ w^m/m! = \exp(\Sigma_{n,P} \#M(n,\varepsilon(P))\ w(P)\ z^n/n!)$$

where the left sum ranges over all $n,m\in \mathbb{N}_o(\underset{\sim}{P})$ and the right sum over all $n\in \mathbb{N}_o(\underset{\sim}{P})$, $n\neq 0$, and $P\in\underset{\sim}{P}$. In particular,

$$\Sigma_n \#Qu_M(X(n))\ z^n/n! = \exp(\Sigma_{n\neq 0}\ iq(n)\ z^n/n!) \quad .$$

Proof: From (3.4) we know that the category $\underline{H}$ is a KS-category with finite automorphism groups and unique KS-partitions. We define a faithful functor $F:\underline{H}\to\underline{K}\times\underline{K}$ by mapping an object $s:X\to Y$ to (X,Y) and a morphism (c,d) to itself. Then both conditions of chapter II, §2, are satisfied. In the product category $\underline{K}\times\underline{K}$ a system of representatives of indecomposable objects modulo isomorphism is $\{(P,0);P\in\underset{\sim}{P}\}\dot{\cup}\{(0,P);P\in\underset{\sim}{P}\}$, and KS-types correspond to pairs $(n,m)\in \mathbb{N}_o(\underset{\sim}{P})\times\mathbb{N}_o(\underset{\sim}{P})$. So, applying (2.26), we obtain the first equation. To prove the second, put $w(P)= 1/a(\varepsilon(P))$ for all P, and observe that $a(m)= a^m m!$ by (2.18). □

(3.40) COROLLARY: For any $m\in \mathbb{N}_o(\underset{\sim}{P})$,

$$\Sigma_n \#M(n,m)\ z^n/n! = \Pi_P\ (\Sigma_n\ \#M(n,\varepsilon(P))\ z^n/n!\ )^{m(P)} \quad . \ \square$$

§4. Power series representations

Let $\underline{K}$,M,~ be a sheaflike categorical structure, and let Mu: $\underline{Al}_k \to \underline{Mo}$, $R \to Mu(R)$, be the affine monoid of multiplicative functions. In this section we are concerned with faithful power series representations of Mu. We use the notation of the appendix, §4, and work over a ground ring k of characteristic 0.

(3.41) When studying the posets of subobjects or quotient objects in a combinatorial category by a sheaflike categorical structure, it is desirable to realize the monoid Mu as a monoid of transformations in a power series algebra, because such a realization connects the count of subobjects or quotient objects with the composition of special power series. In order to make this idea more precise, suppose that we have a faithful representation

$$\rho_R: Mu(R) \to End(P_R) \ , \ R\in Al_k \ ,$$

of the monoid Mu on a power series algebra $P_k= k[[z_j;j\in J]]$, and recall that the zeta, Möbius, characteristic function ζ,μ,χ and the function $D^{\dim}\zeta$ are all multiplicative. Then μ can be calculated simply by inverting the automorphism $\rho(\zeta)$. From $\rho(\zeta^2) = \rho(\zeta)\circ\rho(\zeta)$ one obtains generating functions for the cardinalities of intervals in $Sub_M(Y)$ or $Qu_M(X)$, according to (1.37) or (1.44). Similarly, generating functions for the polynomials $\chi(u)$ and $D^{\dim}\zeta(u)$, $u\in U$, can be derived. This algebraic method to compute the Möbius function, cardinalities of intervals, characteristic polynomial or Whitney numbers of the posets of subobjects or quotient objects will constantly be applied in subsequent examples. Besides, an application of power series representations to numerical mathematics is given in chapter IV, §3.

(3.42) In this connection the following question arises:

Does there always exist a faithful power series representation of the affine monoid of multiplicative functions ?

At the present stage we cannot give a positive answer although the examples suggest it. An analogous question for Lie algebras, but under different circumstances, is settled in [Go,72]. However, from Theorem (3.25) and (A.28) we know how a representation of the group E(Mu) and hence of the monoid Mu on power series algebras in finitely many indeterminates must look like, up to a change of coordinates.

(3.43) THEOREM: Suppose that k is a field. For any commutative k-algebra R let $P_R = R[[z_1,..,z_r]]$ be the power series algebra in the variables $z_1,..,z_r$ with coefficients in R, and let $P_R(2)$ denote the ideal of power series of order ≥ 2.
Then every power series representation $\rho:Mu \to End(P)$ is equivalent to a power series representation $\sigma:Mu \to End(P)$ where, for $i=1,..,r$,

$$\sigma_R(f)(z_i) \in Rz_1+..+Rz_i+P_R(2) \qquad \text{for all } f\in Mu(R) ,$$

in particular,

$$\sigma_R(f)(z_i) \in Rz_i \qquad \text{if } f\in Mu'(R) ,$$

and

$$\sigma_R(f)(z_i)-z_i \in Rz_1+..+Rz_{i-1}+P_R(2) \qquad \text{if } f\in Mu''(R) . \square$$

(3.44) In the rest of this section we consider a very special situation which appears in some examples later on. Assume that ~ is the isomorphism relation on M and that two indecomposable morphisms in M are isomorphic if and only if both their domains and their codomains are isomorphic.
Let $\underset{\sim}{P}$ be a system of representatives of indecomposable objects in $\underset{\sim}{K}$ modulo isomorphism. Then a bijection between the set U of indecomposable types and a subset S of $\mathbb{N}_0(\underset{\sim}{P})\times\underset{\sim}{P}$ is given by $\overline{s:X\to Y} \to (n,P)$ where n is the Krull-Schmidt type of X and P is the representative of the

isomorphism class of Y. This bijection extends to a monoid isomorphism between the set of types T and $\mathbb{N}_o(S)$, and we identify T with $\mathbb{N}_o(S)$. For $P\in\underset{\sim}{P}$ we put $S(P)=\{n\in\mathbb{N}_o(\underset{\sim}{P});(n,P)\in S\}$, and define linear maps $d,c\colon T\to\mathbb{N}_o(\underset{\sim}{P})$ by

$$d(t)=\Sigma_{(n,P)\in S}\,t(n,P)n \quad\text{and}\quad c(t)=\Sigma_{(n,P)\in S}\,t(n,P)\varepsilon(P)$$

for $t=\Sigma_{(n,P)\in S}\,t(n,P)\varepsilon(n,P)$. Then, if $s:X\to Y$ is a morphism in M of type t, in any KS-decomposition of s there are exactly $t(n,P)$ direct summands whose domains have KS-type n and whose codomains are isomorphic to P. Thus $d(t)$ is the KS-type of X whereas $c(t)$ is the KS-type of Y. From (3.22) we know that the multiplication in the affine monoid of multiplicative functions on T is determined by the elementary section coefficients $G(u;t_1,t_2)$, $u\in U$, $t_1,t_2\in T$. But here $G(\varepsilon(n,P);t_1,t_2)=0$ unless $d(t_1)=n$, $c(t_1)\in S(P)$ and $t_2=\varepsilon(c(t_1),P)$, and only the numbers

(3.45) $\quad G(\varepsilon(n,P);t,\varepsilon(c(t),P))\quad$ where $(n,P)\in S$, $t\in T$ satisfy $d(t)=n$ and $c(t)\in S(P)$,

are interesting.

(3.46) PROPOSITION: Suppose that M is a class of monomorphisms with (S2), that all sets $\mathrm{Sub}_M(Y)$ are finite and that there exists a multiplicative function $\varphi:T\to E(k)$ such that

$$G(\varepsilon(n,P);t,\varepsilon(c(t),P)) = \frac{c(t)!\ \varphi(\varepsilon(n,P))}{t!\ \varphi(t)\ \varphi(\varepsilon(c(t),P))}$$

for all $(n,P)\in S$, $t\in T$ with $d(t)=n$ and $c(t)\in S(P)$ (here $E(k)$ = the group of invertible elements in k, $c(t)!=\Pi_{P\in\underset{\sim}{P}}\,c(t)(P)!$ and $t!=\Pi_{(n,P)\in S}\,t(n,P)!$).

Then the affine monoid of multiplicative functions on T has a faithful representation by endomorphisms of the polynomial algebra $k[z(P);P\in\underset{\sim}{P}]$

$$\rho_R\colon \mathrm{Mu}(R) \to \mathrm{End}(R[z(P);P\in\underset{\sim}{P}]),\quad R\in\mathrm{Al}_k,$$
$$f \to [z(P)\to\Sigma_{n\in S(P)}\,f(\varepsilon(n,P))\,z^n/\varphi(\varepsilon(n,P))] .$$

(3.47) REMARK: In some examples of chapter IV, this representation induces a faithful representation of Mu on the power series algebra $k[[w,z(P);P\in\underset{\sim}{P}]]$ where w is a new indeterminate (similar to the passage from affine to projective space in algebraic geometry).

Proof of Proposition (3.46): As $\mathrm{Sub}_M(P)$ is finite for all $P\in\underset{\sim}{P}$, also $S(P)$ is finite and so $\Sigma_{n\in S(P)}\,f(\varepsilon(n,P))\,z^n/\varphi(\varepsilon(n,P))$ is a polynomial. Clearly $\rho_R(\delta)=[z(P)\to z(P)]$ because $G(\varepsilon(n,P);\varepsilon(n,P),\varepsilon(\varepsilon(P),P))=1$ implies $\varphi(\varepsilon(\varepsilon(P),P))=1$. It remains to show that ρ_R is multiplicative.

Let $f,g\in Mu(R)$ and define $f^*,g^*\in Mu(R)$ by
$f^*(t)=f(t)/\varphi(t)$ and $g^*(t)=g(t)/\varphi(t)$ for all $t\in T$. Then
$\rho_R(f)\rho_R(g)(z(P)) = \rho_R(f)(\Sigma_{n\in S(P)} g^*(\varepsilon(n,P))z^n) =$
$= \Sigma_{n\in S(P)} g^*(\varepsilon(n,P)) \Pi_{Q\in \underset{\sim}{P}} (\Sigma_{m\in S(Q)} f^*(\varepsilon(m,Q))z^m)^{n(Q)}$. But
$(\Sigma_{m\in S(Q)} f^*(\varepsilon(m,Q))z^m)^{n(Q)} = \Sigma\; n(Q)!\; \Pi_{m\in S(Q)} \; (f^*(\varepsilon(m,Q))z^m)^{\lambda_Q(m)}/\lambda_Q(m)!$
where the sum runs over all $\lambda_Q\in \mathbb{N}_o(S(Q))$ such that $\Sigma_{m\in S(Q)}\lambda_Q(m) = n(Q)$.
Multiplying out and observing that
$T= \oplus_{Q\in \underset{\sim}{P}} \mathbb{N}_o(S(Q)\times\{Q\})$, $t= \Sigma_{Q\in \underset{\sim}{P}} \Sigma_{m\in S(Q)} \lambda_Q(m)\varepsilon(m,Q)$, we find that
$\Pi_{Q\in \underset{\sim}{P}} (\Sigma_{m\in S(Q)} f^*(\varepsilon(m,Q))z^m)^{n(Q)} = \Sigma_{t\in T, c(t)=n} \frac{n!}{t!} f^*(t)z^{d(t)}$
by the definition of $c(t)$ and $d(t)$. Consequently,
$\rho_R(f)\rho_R(g)(z(P)) = \Sigma_{t\in T, c(t)\in S(P)} \frac{c(t)!}{t!} f^*(t)g^*(\varepsilon(c(t),P))z^{d(t)}$.
Since $c(t)\in S(P)$ implies that $d(t)\in S(P)$, we have
$\rho_R(f)\rho_R(g)(z(P)) = \Sigma_{n\in S(P)} \;(..)\; z^n/\varphi(\varepsilon(n,P))$ where
$(..) = \Sigma_{t\in T, d(t)=n, c(t)\in S(P)} \frac{c(t)!\;\varphi(\varepsilon(n,P))}{t!\;\varphi(t)\;\varphi(\varepsilon(c(t),P))} f(t)g(\varepsilon(c(t),P)) =$
$= \Sigma_{t_1,t_2\in T} G(\varepsilon(n,P);t_1,t_2)f(t_1)g(t_2) = fg(\varepsilon(n,P))$.
Thus $\rho_R(f)\rho_R(g)(z(P)) = \rho_R(fg)(z(P))$ as desired. □

(3.48) PROPOSITION: Suppose that M is a class of epimorphisms with (Q2), that all sets $Qu_M(X)$ are finite and that there exists a multiplicative function $\varphi: T\to E(k)$ such that

$$G(\varepsilon(n,P);t,\varepsilon(c(t),P)) = \frac{c(t)!\;\varphi(\varepsilon(n,P))}{t!\;\varphi(t)\;\varphi(\varepsilon(c(t),P))}$$

for all $(n,P)\in S$, $t\in T$ with $d(t)=n$ and $c(t)\in S(P)$.
Then the affine monoid of multiplicative functions on T has the faithful power series representation

$$\rho_R: Mu(R) \to End(R[[z(P);P\in \underset{\sim}{P}]]) \quad , R\in Al_k ,$$
$$f \to [z(P) \to \Sigma_{n\in S(P)} f(\varepsilon(n,P))\; z^n/\varphi(\varepsilon(n,P))] \quad .$$

Proof: We first show that for every $f\in Mu(R)$ there is a unique endomorphism $\rho_R(f)$ of $P_R = R[[z(P);P\in \underset{\sim}{P}]]$ such that
$\rho_R(f)(z(P)) = \Sigma_{n\in S(P)} f(\varepsilon(n,P))z^n/\varphi(\varepsilon(n,P))$ for all $P\in \underset{\sim}{P}$.
Define $f^*\in Mu(R)$ by $f^*(t)=f(t)/\varphi(t)$. Since $S(P)$ does not contain 0, $w(P) = \Sigma_{n\in S(P)} f^*(\varepsilon(n,P))z^n$ lies in the ideal $P_R(1)$ of all formal power series with zero constant term. For every $n\in \mathbb{N}_o(\underset{\sim}{P})$, the same calculation as in the proof of (3.46) yields
$w^n = \Pi_{Q\in \underset{\sim}{P}} w(Q)^{n(Q)} = \Sigma_{t\in T, c(t)=n} \frac{n!}{t!} f^*(t)z^{d(t)}$.
In particular, the coefficient of z^m, $m\in \mathbb{N}_o(\underset{\sim}{P})$, is

$\Sigma_{t\in T, d(t)=m, c(t)=n} \frac{n!}{t!} f^*(t)$.

But the set $\{t\in T; d(t)=m\}$ is finite because the sets $Qu_M(X)$ are finite, and hence $(w^n; n\in \mathbb{N}_o(\underset{\sim}{P}))$ is a 0-family. Therefore we can define a continuous R-algebra homomorphism $\rho_R(f): P_R \to P_R$ by

$\rho_R(f)(\Sigma_n \lambda^n z^n) = \Sigma_m (\Sigma_{t\in T, d(t)=m} \frac{c(t)!}{t!} \lambda^{c(t)} f^*(t))\ z^m$.

This clearly is the unique endomorphism of P_R mapping each $z(P)$ to $w(P)$. Now it can be proved as in (3.46) that ρ is a faithful representation.

□

§5. The basic examples

To illustrate the formalism developed so far, we now discuss two examples: the Boolean algebras of subsets of finite sets and the geometric lattices of partitions of finite sets. Of course, we prove nothing new and only want to show how the theory works, but the method of computation used here will also be applied in the new examples of chapter IV.

Let $\underline{K}$ be the category $\underline{Sf}$ of finite sets. In $\underline{Sf}$ a system of representatives of the isomorphism classes of indecomposable objects is given by $\underset{\sim}{P}=\{P\}$ where P is a set with a single element. We identify Krull-Schmidt types with natural numbers by $\mathbb{N}_o(\underset{\sim}{P}) = \mathbb{N}_o$, $n = n(P)$. Then the KS-type of a finite set is its cardinality.

(3.49) Subsets of finite sets:

Let M be the class of monomorphisms in $\underline{Sf}$, i.e. the class of one-to-one maps, and let $\sim$ be the isomorphism relation on M. Then, for every finite set Y, the poset $Sub_M(Y)$ is the power set of Y ordered by inclusion.

Any indecomposable monomorphism is isomorphic either to $\emptyset\to P$ or to $P\to P$. Thus we are in the situation of (3.44) with

$$S=\{(1,P),(0,P)\} \quad \text{and} \quad T= \mathbb{N}_o(S) \ .$$

If the monomorphism $s:X\to Y$ has the type $t\in T$, then $t(1,P)= \#X$ and $t(0,P)= \#Y - \#X$. The elementary section coefficients (3.45) are $G(\varepsilon(0,P);0,\varepsilon(0,P))= G(\varepsilon(0,P);\varepsilon(0,P),\varepsilon(1,P))= G(\varepsilon(1,P);\varepsilon(1,P),\varepsilon(1,P))= 1$ according to (1.33). Applying Proposition (3.46) with $\varphi(\varepsilon(0,P))=$ $= \varphi(\varepsilon(1,P))= 1$, we obtain a faithful representation of the affine monoid of multiplicative functions on T

$$\rho_R: Mu(R) \to End(R[z]) \quad , R\in Al_{\mathbb{Z}} ,$$
$$f \to [z \to f(\varepsilon(1,P))z + f(\varepsilon(0,P))] \quad .$$

In particular, Mu is isomorphic to the monoid of affine transformations on the affine line. As $U'=\{\varepsilon(1,P)\}$ and $U''=\{\varepsilon(0,P)\}$, under this isomorphism the diagonalizable monoid Mu' corresponds to the monoid of dilatations whereas the unipotent group Mu" corresponds to the group of translations which is isomorphic to the additive group G_a , $G_a(R) = R$ with +.

Since $\zeta \to [z \to z+1]$

and $D^{dim} \to [z \to z+D]$,

it follows that $\mu \to [z \to z-1]$,

$\chi=\mu D^{dim} \to [z \to z-1+D]$

and $D^{dim}\zeta \to [z \to z+D+1]$.

So, by Theorem (1.37), the characteristic polynomial of the power set of $\{1,..,n\}$ is

$$\chi(n\varepsilon(0,P)) = \chi(\varepsilon(0,P))^n = (D-1)^n$$

because the inclusion map $\emptyset \to \{1,..,n\}$ has type $n\varepsilon(0,P)$.

The generating polynomial of the Whitney numbers is

$$D^{dim}\zeta(n\varepsilon(0,P)) = (D+1)^n = \Sigma_{i=0}^{n}\binom{n}{i}D^i \quad ,$$

and the Whitney numbers $W(i)$ are the binomial coefficients $\binom{n}{i}$.

A faithful power series representation of Mu is given by

$$\hat{\rho}_R: Mu(R) \to End(R[[w,z]]) \quad , R\in Al_{\mathbb{Z}} ,$$
$$f \to \begin{bmatrix} w \to w \\ z \to f(\varepsilon(1,P))z + f(\varepsilon(0,P))w \end{bmatrix} \quad .$$

We conclude with some remarks. Instead of the isomorphism relation also the equivalence ~ from (1.53) can be chosen. In this case the affine monoid Mu is isomorphic to the additive group G_a . Hence, by (3.32), the affine monoid of multiplicative functions of the product structure $\underline{K}^S, M^S, \sim^S$ is isomorphic to the product group $(G_a)^S$ whose affine algebra is the "binomial coalgebra" of S.A.Joni and G.C.Rota in [JR,79].

(3.50) <u>Partitions of finite sets:</u>

Let M be the class of epimorphisms in <u>Sf</u> , i.e. the class of surjective maps, and let ~ be the isomorphism relation on M.

A quotient object $[r:X \to Y]$ of a finite set X usually is identified with the partition $\{r^{-1}(y); y\in Y\}$ of X, and then the poset $Qu_M(X)$ is the set of partitions of X ordered by refinement. The cardinalities of the sets $Qu_M(X)$ are called the Bell numbers $B(n)$. By (3.39) their generating function is

$$\Sigma_{n=0}^{\infty} B(n)\, z^n/n! = \exp(e^z-1) \quad .$$

The Stirling numbers of the second kind $S(n,m)$ count the partitions of $\{1,..,n\}$ with m blocks. Since $S(n,m) = \#M(n,m)/m!$ by (3.38), (3.39) implies that

$$\Sigma_{n,m} S(n,m)\ w^m\ z^n/n! = \exp(w(e^z-1)) \quad .$$

Any indecomposable epimorphism is isomorphic to exactly one of the surjections $nP \to P$, $n=1,2,..$. Again we are in the situation of (3.44) with

$$S=\{(n,P);n=1,2,..\} \quad \text{and} \quad T= \mathbb{N}_o(S) \quad .$$

If the epimorphism $s:X \to Y$ has the type $t \in T$, then $t(n,P)$ is the number of blocks of the partition $\{s^{-1}(y);y \in Y\}$ of size n. The elementary section coefficients (3.45) are

$$G(\varepsilon(n,P);t,\varepsilon(c(t),P)) = n!/\ \Pi_{i=1}^{\infty}(i!)^{t(i,P)}t(i,P)! \qquad (\ \Pi_{i=1}^{\infty}t(i,P)i = n\ )$$

according to (1.40) and (2.8). Applying Proposition (3.48) with $\varphi(\varepsilon(n,P)) = n!$, we get a faithful power series representation of the affine monoid of multiplicative functions on T

$$\rho_R: Mu(R) \to End(R[[z]]) \quad , \quad R \in Al_{\mathbb{Q}} \ ,$$
$$f \to [z \to \Sigma_{n=1}^{\infty}\ f(\varepsilon(n,P))\ z^n/n!] \quad .$$

Hence, over the rational numbers, Mu is isomorphic to the endomorphism monoid of the power series algebra in one indeterminate. As $U'=\{\varepsilon(1,P)\}$ and $U''=\{\varepsilon(n,P);n \geq 2\}$, under this isomorphism the diagonalizable monoid Mu' corresponds to the monoid of diagonal endomorphisms whereas the unipotent group Mu" corresponds to the group of all automorphisms such that $z \to z$ + terms of order ≥ 2 .

The dimension of an epimorphism $s:X \to Y$ is $\#Y - \#X$. Therefore, $\dim: T \to \mathbb{N}_o$ is given by $\dim(\varepsilon(n,P)) = n-1$. Under the isomorphism ρ the normal subgroups Mu"(d) of Mu" from (3.26) correspond to the groups of all automorphisms such that $z \to z$ + terms of order $\geq d+1$. Here the sets $U(1,d) = \{\varepsilon(n,P);2 \leq n \leq d\}$ are finite and so the groups Mu"/Mu"(d) are algebraic. An easy computation shows that

$$G''(\varepsilon(n,P);\varepsilon(n_1,P),\varepsilon(n_2,P)) = \binom{n}{n_1} \quad \text{if } n+1= n_1+n_2 \ , \text{ and } \quad =0 \quad \text{else} \ .$$

By (3.28) the Lie algebra of Mu" has the topological basis

$$b(n) = e''(\varepsilon(n,P)) \quad , \quad n=2,3,.. \quad ,$$

with the Lie bracket

$$[b(n_1),b(n_2)] = \{\binom{n_1+n_2-1}{n_1} - \binom{n_1+n_2-1}{n_2}\}\ b(n_1+n_2-1) \quad .$$

Since $\zeta \to [z \to \exp(z)-1]$

and $D^{\dim} \to [z \to (e^{Dz}-1)/D]$,

it follows that $\mu \to [z \to \log(1+z)]$,

$\chi=\mu D^{\dim} \to [z \to ((1+z)^D-1)/D]$

and $D^{\dim}\zeta \to [z \to \exp((e^{Dz}-1)/D]$.

Expanding into power series, we read off that

$$\mu(\varepsilon(n,P)) = (-1)^{n-1}(n-1)! \quad ,$$

$$\chi(\varepsilon(n,P)) = (D-1)(D-2)..(D-n)$$

and
$$D^{\dim}\zeta(\varepsilon(n,P)) = \Sigma_{i=0}^{n-1} S(n,n-i)D^i \quad .$$

Now consider the lattice Π_n of partitions of $\{1,..,n\}$ which has the least element $\hat{0}=\{\{1\},..,\{n\}\}$ and the greatest element $\hat{1}=\{\{1,..,n\}\}$. Since the canonical map $\{\{1\},..,\{n\}\}\to\{\{1,..,n\}\}$ has type $\varepsilon(n,P)$, Theorem (1.44) implies that the Möbius function on Π_n takes the value

$$\mu_{\Pi_n}(\hat{0},\hat{1}) = (-1)^{n-1}(n-1)! \quad ,$$

that the characteristic polynomial of Π_n is

$$(D-1)(D-2)..(D-n) \quad ,$$

and that the Whitney numbers $W(i)$ of Π_n are the $S(n,n-i)$.

The affine algebra of Mu is called the Faà di Bruno bialgebra in [JR,79] and studied in [Dou,74]. The power series representation ρ is due to P.Doubilet, G.C.Rota and R.Stanley ([DRS,72]). The Möbius function on the lattice of partitions of a finite set was first determined by M.P.Schützenberger ([Sch,54]), and was applied to symmetric functions in [Dou,72].

§6. Auxiliary considerations

Let $\underline{K}$,M,~ be a sheaflike categorical structure where ~ is the isomorphism relation on M. In this section we derive a formula for the number of morphisms $s:X\to Y$ in M of type t, which will be used in the next chapter to compute elementary section coefficients.

Let $\underset{\sim}{P}$ be a system of representatives of the isomorphism classes of indecomposable objects in $\underline{K}$. We define linear maps $d,c: T \to \mathbb{N}_o(\underset{\sim}{P})$ by

$$d(t) = \text{KS-type of } X \quad \text{and} \quad c(t) = \text{KS-type of } Y$$

where $s:X\to Y$ is an arbitrary representative of t. Let S be the image of the map $U \to \mathbb{N}_o(\underset{\sim}{P})\times\underset{\sim}{P}$, $\overline{s:X\to Y} \to (n,P)$ where n is the KS-type of X and P is the representative of the isomorphism class of Y. Extending linearly we get a surjective homomorphism of monoids

$$\nu: T \to \mathbb{N}_o(S) \quad .$$

Then, if $s:X\to Y$ is a morphism in M of type t, in any KS-decomposition of s there are exactly $\nu(t)(n,P)$ direct summands whose domains have KS-type n and whose codomains are isomorphic to P. Consequently,

$$d(t) = \Sigma_{(n,P)\in S}\, \nu(t)(n,P)n \quad \text{and} \quad c(t) = \Sigma_{(n,P)\in S}\, \nu(t)(n,P)\varepsilon(P) \quad .$$

Now, for $m \in \mathbb{N}_o(\underset{\sim}{P})$ let $\qquad X(m) = \oplus_P m(P)P \quad ,$

and for $t \in T$ let $\qquad M(t) = \{s : X(d(t)) \to X(c(t))\,;\, \bar{s} = t\} \quad .$

(3.51) PROPOSITION: Suppose that all sets $M(u)$, $u \in U$, are finite. Then for any type $t = \Sigma_{u \in U} t(u)u$, the set $M(t)$ is finite and

$$\#M(t) = \frac{d(t)!\ c(t)!}{t!\ N(\nu(t))}\ \Pi_{u \in U} \#M(u)^{t(u)}$$

where $N(\nu(t)) = \Pi_{(m,Q) \in S} (m!)^{\nu(t)(m,Q)}$.

(3.52) COROLLARY: Suppose that M consists of epimorphisms and that all sets $M(u)$, $u \in U$, are finite. Then for any $n \in \mathbb{N}_o(\underset{\sim}{P})$ and $t \in T$ with $d(t) = n$, the number of quotient objects $[r]$ of $X(n)$ such that r has type t is

$$\frac{n!}{t!\ E(\nu(t))}\ \Pi_{u \in U} \#M(u)^{t(u)}$$

where $E(\nu(t)) = \Pi_{(m,Q) \in S} (a(Q)m!)^{\nu(t)(m,Q)}$.

Proof: The number in question is $\#M(t)/a(c(t))$ by (3.38), and $a(c(t)) = \Pi_{Q \in \underset{\sim}{P}}\ a(Q)^{c(t)(Q)} c(t)(Q)! = (\Pi_{(m,Q) \in S} a(Q)^{\nu(t)(m,Q)})\ c(t)!$. □

Proof of Proposition (3.51): We first introduce some notation. Let $m \in \mathbb{N}_o(\underset{\sim}{P})$. For every $Q \in \underset{\sim}{P}$ we label the $m(Q)$ canonical injections $Q \to \oplus_Q m(Q)Q = X(m)$ by $(Q,1),\ldots,(Q,m(Q))$ and write $Q \xrightarrow{(Q,i)} X(m)$.

Let $\quad [m] = \cup_{Q \in \underset{\sim}{P}} \{Q\} \times \{1,\ldots,m(Q)\} \quad$ which is a finite set.

For any subset $I = \cup_{Q \in \underset{\sim}{P}} \{Q\} \times I_Q$ of $[m]$, let $\oplus_Q (\#I_Q)Q \xrightarrow{I} X(m)$ be the monomorphism induced by the injections $Q \xrightarrow{(Q,i)} X(m)$, $(Q,i) \in I$. Finally, for $n, q \in \mathbb{N}_o(\underset{\sim}{P})$ let $M(n,q) = \{s : X(n) \to X(q)\,;\, s \in M\}$. Then we have the following result.

(3.53) LEMMA: Let $n, q \in \mathbb{N}_o(\underset{\sim}{P})$. For any map $A : [n] \to [q]$, define $Z_A \in \mathbb{N}_o(\underset{\sim}{P})^{[q]}$ by $Z_A(Q,i)(P) =$ number of $j \in \{1,\ldots,q(P)\}$ such that $A(P,j) = (Q,i)$.

Then the map

$$\Psi : \cup_{A:[n]\to[q]} \{A\} \times \Pi_{(Q,i) \in [q]} M(Z_A(Q,i), \varepsilon(Q)) \to M(n,q)$$
$$(A, (s_{Q,i})) \to s \quad ,$$

defined by the commutative diagrams

$$\begin{array}{ccc} X(n) & \xrightarrow{\quad s \quad} & X(q) \\ {\scriptstyle A^{-1}(Q,i)} \uparrow & & \uparrow {\scriptstyle (Q,i)} \\ X(Z_A(Q,i)) & \xrightarrow{\quad s_{Q,i} \quad} & P \end{array} \qquad , \qquad (Q,i) \in [q] \quad ,$$

is a bijection.

Proof: By (3.2) Ψ is one-to-one. To show that Ψ is onto, let $s\in M(n,q)$. Then the KS-partition of s yields a partition π of $X(n)$ which is indexed by not necessarily all $(Q,i)\in[q]$. By the uniqueness of KS-partitions in $\underline{K}$, we have $\pi = \{[\oplus_P(\# I(Q,i)_P)P \xrightarrow{I(Q,i)} X(n)]\}$ where the subsets $I(Q,i)$ form a partition of the set $[n]$. From the commuting diagrams

$$\begin{array}{ccc} X(n) & \xrightarrow{\quad s \quad} & X(q) \\ {\scriptstyle I(Q,i)}\uparrow & & \uparrow{\scriptstyle (Q,i)} \\ \oplus_P(\# I(Q,i)_P)P & \xrightarrow{\quad s_{Q,i} \quad} & Q \end{array}$$

we conclude that $s = \Psi(A,(s_{Q,i}))$ where $A:[n]\to[q]$ is given by $A(P,j)=(Q,i)$ for $(P,j)\in I(Q,i)$. □

After these preparations we prove (3.51). Let $n=d(t)$ and $q=c(t)$. Under the bijection Ψ an $s\in M(t)$ corresponds to an $(A,(s_{Q,i}))$ where, for every $(m,Q)\in S$, the number of $i\in\{1,..,q(Q)\}$ such that $Z_A(Q,i)=m$ is $\nu(t)(m,Q)$, and where $\Sigma_{(Q,i)\in[q]}\ \bar{s}_{Q,i} = t$.

But for each such A we have a bijection

$$\Pi_{(Q,i)\in[q]}\ M(Z_A(Q,i),\varepsilon(Q)) \to \Pi_{(m,Q)\in S}\ M(m,\varepsilon(Q))^{\nu(t)(m,Q)}$$

$$(s_{Q,i}) \to (s_{(m,Q),j}) \quad ,$$

and the number of $(s_{(m,Q),j})$ with $\Sigma_{(m,Q)\in S}\ \Sigma_{j=1}^{\nu(t)(m,Q)}\ \bar{s}_{(m,Q),j} = t$ is $\frac{\nu(t)!}{t!}\ \Pi_{u\in U}\# M(u)^{t(u)}$ by comparing coefficients in the basis U.

It remains to determine the number of maps $A:[n]\to[q]$ with the following property: For every $(m,Q)\in S$, there are exactly $\nu(t)(m,Q)$ elements i in $\{1,..,q(Q)\}$ such that $Z_A(Q,i)=m$. We denote the set of such maps by $A(t)$. Let $G(q)$ be the group of all permutations of $[q]$ leaving each set $\{Q\}\times\{1,..,q(Q)\}$, $Q\in\underset{\sim}{P}$, stable. Obviously, $G(q)$ is isomorphic to a product of symmetric groups $\Pi_{Q\in\underset{\sim}{P}}\ S_{q(Q)}$ and has the order $q!$.

Let $A(n,q)$ be the set of all maps $A:[n]\to[q]$. Then, under the operation $G(n)\times G(q)\times A(n,q) \to A(n,q)$, $(g,h,A) \to hAg^{-1}$, the set $A(t)$ is stable. But it is not difficult to see that $G(n)\times G(q)$ acts transitively on $A(t)$, and that the stabilizer of a map $A\in A(t)$ is isomorphic to a direct product of wreath products $\Pi_{(m,Q)\in S}\ (\Pi_{P\in\underset{\sim}{P}}\ S_{m(P)})\ \mathrm{wr}\ S_{\nu(t)(m,Q)}$ and has the order $N(\nu(t))\nu(t)!$.

Therefore, $\# A(t) = \dfrac{n!\ q!}{N(\nu(t))\ \nu(t)!}$ which ends the proof. □

CHAPTER IV

APPLICATIONS

In this chapter we study several sheaflike categorical structures of combinatorial interest. We will meet various kinds of unipotent groups, and often will establish a faithful power series representation of the affine monoid of multiplicative functions. In all examples the conditions of (3.11) are easily verified (In order to look them up, use the subject index).

§1. Systems of finite sets

Let J be a finite partially ordered set. A J-system of finite sets is an order-preserving map $J\to Pot(X)$, $j\to X_j$, where X is a finite set and Pot(X) is the power set of X ordered by inclusion. We briefly denote it by $(X,(X_j))$. Given two J-systems $(X,(X_j))$ and $(Y,(Y_j))$, a morphism from $(X,(X_j))$ to $(Y,(Y_j))$ is a map $s:X\to Y$ such that $s(X_j)\subset Y_j$ for all $j\in J$. Let J-Sy be the category of J-systems of finite sets. If J is discrete, i.e. if any two elements of J are incomparable, then J-Sy is the category of finite sets with a family of subsets indexed by J (These are the "systems" of [GW,77]). If J is totally ordered, then J-Sy is the category of finite sets with an ascending sequence of subsets indexed by J. Finally, if J is empty, then J-Sy reduces to the category of finite sets.
Let $(Y,(Y_j))$ be a J-system. A J-subsystem of $(Y,(Y_j))$ is a J-system $(Z,(Z_j))$ where $Z\subset Y$ and $Z_j\subset Y_j$ for all j. Since a morphism $s:(X,(X_j))\to(Y,(Y_j))$ is a monomorphism if and only if $s:X\to Y$ is one-to-one, we can identify the subobject $[r:(X,(X_j))\to(Y,(Y_j))]$ of $(Y,(Y_j))$ with the J-subsystem $(r(X),(r(X_j))$. Then $(W,(W_j))\leq(Z,(Z_j))$ in $Sub(Y,(Y_j))$ if and only if $W\subset Z$ and $W_j\subset Z_j$ for all j. Obviously $Sub(Y,(Y_j))$ is a lattice.
In J-Sy the only initial object 0 is the empty J-system $(\emptyset,(\emptyset))$, and the direct sum of two J-systems $(X,(X_j))$ and $(Y,(Y_j))$ is the disjoint union $(X\dot{\cup}Y,(X_j\dot{\cup}Y_j))$. A J-system $(W,(W_j))$ is indecomposable if and only

if W has exactly one element. In this case the set $C=\{j\in J; W_j=W\}$ is a coideal of J, i.e., for any $c\in C$ and $j\in J$, $c\leq j$ implies that $j\in C$. A morphism $s:(X,(X_j))\to(Y,(Y_j))$ is an isomorphism if and only if $s:X\to Y$ is a bijection and $s(X_j)=Y_j$ for all j. Hence a system of representatives of the isomorphism classes of indecomposable objects in J-Sy is

$$\underset{\sim}{P} = \{P_C;\ C \text{ a coideal of } J\}$$

where $P_C=(\{.\},(\{.\} \text{ if } j\in C,\ \emptyset \text{ if } j\notin C))$ and $\{.\}$ is a set with one element. Let Co(J) denote the finite set of coideals of J. A J-system $(X,(X_j))$ has the unique Krull-Schmidt partition $\{(\{x\},(\{x\}\cap X_j));x\in X\}$. Hence the KS-type $n\in \mathbb{N}_o(\underset{\sim}{P})$ of $(X,(X_j))$ is given by

$$n(P_C) = \#\{x\in X;\ x\in X_j \text{ for all } j\in C, \text{ and } x\notin X_j \text{ for all } j\notin C\}\ ,\quad C\in Co(J)\ .$$

For instance, if J is the discrete poset {1,2,3}, then Co(J) is the power set of {1,2,3} and one can read off the KS-type of $(X,(X_1,X_2,X_3))$ from the diagram below.

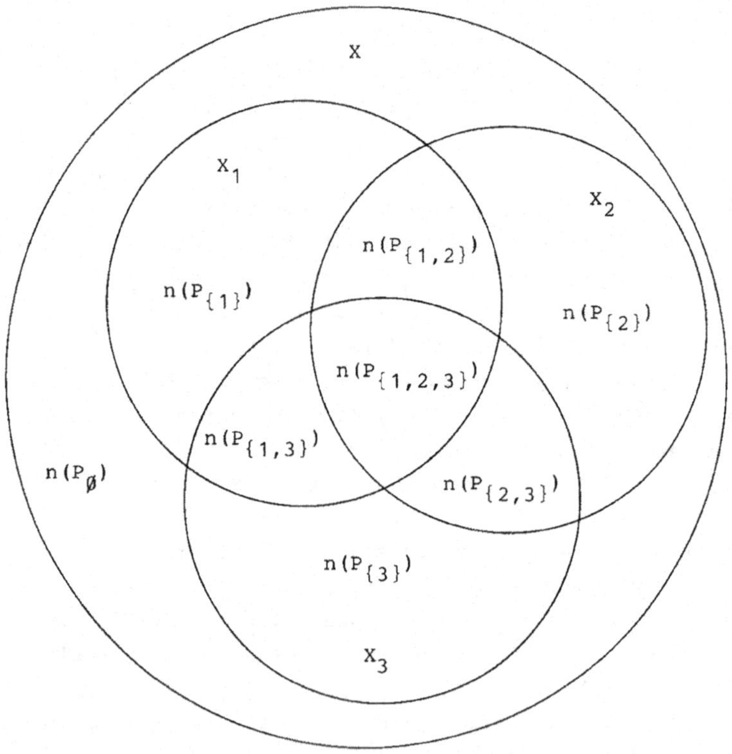

The order of the automorphism group of an object $(X,(X_j))$ of KS-type n is, by (2.18),

$$a(n) = \Pi_{C\in Co(J)}\ n(P_C)! = n!\ .$$

In the sequel we determine the affine monoids of multiplicative functions for the posets of subobjects and quotient objects in J-Sy. To simplify the notation, we write C instead of P_C when no confusion is possible.

First case:

Let M be the class of monomorphisms, and let ~ denote the isomorphism relation on M.
Any indecomposable monomorphism in J-Sy is isomorphic to exactly one of the morphisms $0 \to P_C$, $C \in Co(J)$, or $P_B \to P_C$, $B,C \in Co(J)$ such that $B \subset C$. Thus we are in the situation of (3.44) with
$S = \{(0,C); C \in Co(J)\} \cup \{(\varepsilon(B),C); B,C \in Co(J) \text{ such that } B \subset C\}$ and $T = \mathbb{N}_0(S)$.
The elementary section coefficients (3.45) are $G(\varepsilon(0,C);0,\varepsilon(0,C)) =$
$= G(\varepsilon(0,C);\varepsilon(0,B),\varepsilon(\varepsilon(B),C)) = G(\varepsilon(\varepsilon(A),C);\varepsilon(\varepsilon(A),B),\varepsilon(\varepsilon(B),C)) = 1$
according to (1.33). Applying Proposition (3.46) with $\varphi(\varepsilon(0,C)) =$
$= \varphi(\varepsilon(\varepsilon(B),C)) = 1$, we obtain the following result.

(4.1) THEOREM: The affine monoid of multiplicative functions on T is algebraic and has the faithful representation

$$\begin{aligned} \rho_R: Mu(R) &\to End(R[z(C); C \in Co(J)]) \qquad , R \in Al_{\mathbb{Z}} , \\ f &\to [z(C) \to f(\varepsilon(0,C)) + \Sigma_{B \subset C} f(\varepsilon(\varepsilon(B),C))z(B)] \end{aligned}$$

by linear endomorphisms of the polynomial algebra $\mathbb{Z}[z(C); C \in Co(J)]$. □

ρ induces the faithful power series representation

$$\begin{aligned} \hat{\rho}_R: Mu(R) &\to End(R[[w, z(C); C \in Co(J)]]) \qquad , R \in Al_{\mathbb{Z}} , \\ f &\to \begin{bmatrix} w \to w \\ z(C) \to f(\varepsilon(0,C))w + \Sigma_{B \subset C} f(\varepsilon(\varepsilon(B),C))z(B) \end{bmatrix} . \end{aligned}$$

To compute the Möbius function $\mu \in Mu''(\mathbb{Z})$, we have to invert
$\rho(\zeta) = [z(C) \to 1 + \Sigma_{B \subset C} z(B)]$. Solving the equations
$1 + \Sigma_{B \subset C}\, \rho(\mu)(z(B)) = z(C)$, $C \in Co(J)$, by Möbius inversion on the set Co(J) ordered by inclusion, it follows that
$\rho(\mu) = [z(C) \to -\delta_{\emptyset,C} + \Sigma_{B \subset C}\, \mu_{Co(J)}(B,C)z(B)]$
where $\delta_{\emptyset,C}$ is the Kronecker delta and $\mu_{Co(J)}$ is the Möbius function of Co(J). Hence

$$\mu(\varepsilon(0,C)) = -1 \text{ if } C = \emptyset \quad \text{and} \quad \mu(\varepsilon(0,C)) = 0 \text{ else ,}$$
$$\text{and} \quad \mu(\varepsilon(\varepsilon(B),C)) = \mu_{Co(J)}(B,C) \quad .$$

Instead of these algebraic arguments, μ could have been determined also by combinatorial means: The lattice $Sub(P_J)$ is isomorphic to the poset Co(J) with a new least element adjoined, and Theorem (1.37) relates the Möbius function of the poset $\mathrm{Sub}(P_J)$ to $\mu \in \mathrm{Mu}''(\mathbb{Z})$.

Second case:

Let M be the class of epimorphisms, and let $\sim$ denote the isomorphism relation on M.
An epimorphism is a morphism $s:(X,(X_j)) \to (Y,(Y_j))$ where $s:X \to Y$ is onto.
If $s:(X,(X_j)) \to P_C$ is an epimorphism, then the KS-type n of $(X,(X_j))$ is non-zero and the support of n, $\mathrm{supp}(n) = \{B \in \mathrm{Co}(J); n(B) \neq 0\}$, is contained in the set of coideals of C.
A quotient object $[r:(X,(X_j)) \to (Y,(Y_j))]$ of a J-system $(X,(X_j))$ can be identified with the J-system $(\{r^{-1}(y); y \in Y\}, (\{r^{-1}(y); y \in Y_j\}))$. Then $\mathrm{Qu}(X,(X_j))$ is the set of all J-systems $(\pi,(\pi_j))$ where π is a partition of X and each X_j is covered by the blocks in π_j. In $\mathrm{Qu}(X,(X_j))$ we have $(\pi,(\pi_j)) \leq (\sigma,(\sigma_j))$ if and only if π is a refinement of σ and each block of π_j is contained in a block of σ_j, for every j. With this order $\mathrm{Qu}(X,(X_j))$ is a lattice.
We next count, given two J-systems $(X,(X_j))$ and $(Y,(Y_j))$, the surjections $s:X \to Y$ such that $s(X_j) \subset Y_j$ for all j.

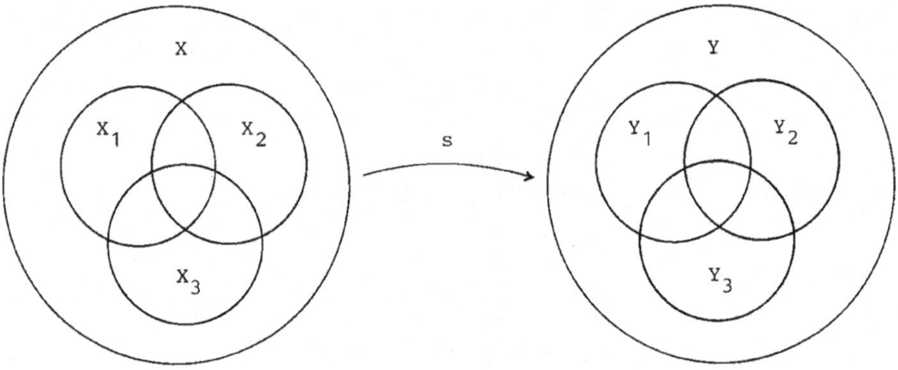

For instance, if X is a set of different balls and Y is a set of different boxes, one can ask for the number of ways how the balls can be distributed into the boxes without leaving any box empty, but observing some restrictions (e.g. red balls in large boxes, small balls in blue boxes,..). Without restrictions, i.e. when J is empty, this is a basic distribution problem in combinatorics whose solution is $\Sigma_{l=0}^{m} (-1)^{m-l}\binom{m}{l} l^n$ where $n = \#X$ and $m = \#Y$ ([Ri,58]). In the general situation, the number in question is $\#M(n,m)$ where n,m are the KS-types of X or Y respectively.

(4.2) PROPOSITION: For $n,m \in \mathbb{N}_o(\underset{\sim}{P})$,

$$\#M(n,m) = \Sigma_1 \Pi_C (-1)^{m(C)-1(C)} \binom{m(C)}{1(C)} (\Sigma_{D\supset C} 1(D))^{n(C)}$$

where the first sum runs over all $1 \in \mathbb{N}_o(\underset{\sim}{P})$ such that $1(C) \leq m(C)$ for every $C \in Co(J)$.

Proof: By (3.40), $\Sigma_n \#M(n,m) z^n/n! = \Pi_C \{\exp(\Sigma_{B\subset C} z(B)) -1\}^{m(C)}$ in $\mathbb{Q}[[z(C);C \in Co(J)]]$ because $\Sigma_n \#M(n,\varepsilon(P_C))z^n/n! = \exp(\Sigma_{B\subset C} z(B)) -1$. Multiplying out and comparing coefficients, we obtain the stated formula. □

In J-Sy any indecomposable epimorphism is isomorphic to exactly one of the morphisms $\oplus_B n(B) P_B \to P_C$ where $n \neq 0$ and $supp(n) \subset Co(C)$. Again we are in the situation of (3.44) with

$$S = \{(n,C); n \in \mathbb{N}_o(\underset{\sim}{P}), C \in Co(J) \text{ such that } n \neq 0 \text{ and } supp(n) \subset Co(C)\}$$
$$\text{and} \quad T = \mathbb{N}_o(S) \quad .$$

To compute the elementary section coefficients (3.45), let $(X,(X_j))$ be a J-system of KS-type n and let $s:(X,(X_j)) \to P_C$. By (1.40) $G(\varepsilon(n,C);t,\varepsilon(c(t),C)) = \#\{[r] \in Qu(X,(X_j));[r] \leq [s], \bar{r}=t, \overline{s/r}=\varepsilon(c(t),C)\}$. But $\bar{r}=t$ implies that $[r] \leq [s]$ and that $\overline{s/r}=\varepsilon(c(t),C)$, so formula (3.52) yields

$$G(\varepsilon(n,C);t,\varepsilon(c(t),C)) = \frac{n!}{t!\ \Pi_{(m,Q)\in S} (m!)^{t(m,Q)}} \quad .$$

Applying Proposition (3.48) with $\varphi(\varepsilon(n,C)) = n!$, we have proved

(4.3) THEOREM: The affine monoid of multiplicative functions on T has the faithful power series representation

$$\rho_R: Mu(R) \to End(R[[z(C);C \in Co(J)]]) \quad , R \in Al_{\mathbb{Q}} \; ,$$
$$f \to [z(C) \to \Sigma_n f(\varepsilon(n,C)) z^n/n!] \quad . \; \square$$

To determine the Möbius function $\mu \in Mu''(\mathbb{Z})$, $\rho(\zeta)$ must be inverted. From the equations $\exp(\Sigma_{B\subset C}\ \rho(\mu)(z(B)) -1 = z(C)$, $C \in Co(J)$, which can be written as $\Sigma_{B\subset C}\ \rho(\mu)(z(B)) = \log(1+z(C))$, $C \in Co(J)$, it follows by Möbius inversion on Co(J) that

$\rho(\mu) = [z(C) \to \Sigma_{B\subset C}\ \mu_{Co(J)}(B,C) \log(1+z(B))]$. Expanding log into a power series, we get the values of μ on the indecomposable types:

$$\mu(\varepsilon(n,C)) = (-1)^{1-1}(1-1)!\ \mu_{Co(J)}(B,C) \quad \text{if } n=1\varepsilon(B) \text{ where } 1 \in \mathbb{N} \text{ and } B \subset C,$$
$$\text{and} \quad = 0 \quad \text{otherwise.}$$

Finally, we use the power series representation ρ to set up a generating function for a counting problem in graph theory.

(4.4) PROPOSITION: Let E be a graphic property (such as connectivity, k-colorability, planarity, etc.). For $n\in\mathbb{N}$ let $C(n,E)$ be the number of connected graphs with vertex set $\{1,..,n\}$ having property E, and let $c_E(z)=\Sigma_{n=1}^{\infty}\, C(n,E)z^n/n!$ be their exponential generating function. For $n,m\in\mathbb{N}_o$ with $n\geq m$, let $N(n,E;m)$ be the number of graphs with vertex set $\{1,..,n\}$ of which exactly m connected components have property E. Then

$$\Sigma_{n=0}^{\infty}\ (\Sigma_{m=0}^{n}\ N(n,E;m)x^m)\ z^n/n! = (\Sigma_{l=0}^{\infty}\ 2^{l(l-1)/2}z^l/l!)\ \exp((x-1)c_E(z))$$

in $\mathbb{Q}[x][[z]]$.

(4.5) COROLLARY: The average number of components with property E of a graph with vertex set $\{1,..,n\}$ is

$$2^{-n(n-1)/2}\ \Sigma_{l=0}^{n}\ \binom{n}{l}2^{l(l-1)/2}C(n-l,E)\quad .$$

Proof: To compute $\bar{N}(n,E)=\Sigma_{m=0}^{n}\ mN(n,E;m)/2^{n(n-1)/2}$, we differentiate in (4.4) with respect to x and afterwards set x=1. This gives $\Sigma_{n=0}^{\infty}\ \bar{N}(n,E)2^{n(n-1)/2}z^n/n! = (\Sigma_{l=0}^{\infty}\ 2^{l(l-1)/2}z^l/l!)c_E(z)$ which implies the result. □

Proof of Proposition (4.4): We shall apply the representation ρ in the special case $J=\{0\}$. Here $Co(J)=\{\emptyset,J\}$, $\mathbb{N}_o(\underset{\sim}{P})=\mathbb{N}_o\times\mathbb{N}_o$ by the identification $n=(n(\emptyset),n(J))$, and $S=\{((k,0),\emptyset);k\in\mathbb{N}\}\cup\{((k,l),J);k,l\in\mathbb{N}_o,k+l>0\}$. Hence, for $R\in Al_{\mathbb{Q}}$,

$$\begin{aligned}\rho_R:\ Mu(R)\ &\to\ End(R[[z(\emptyset),z(J)]])\\ f\ &\to\ \begin{bmatrix} z(\emptyset)\ \to\ \Sigma_k\ f(\varepsilon((k,0),\emptyset))\ z(\emptyset)^k/k! \\ z(J)\ \to\ \Sigma_{k,l}\ f(\varepsilon((k,l),J))\ z(\emptyset)^k z(J)^l/k!l!\end{bmatrix}\ .\end{aligned}$$

Now let $X=\{1,..,n\}$, and consider the lattice $Qu(X,\emptyset)=\{(\pi,\pi_0);\pi$ a partition of X, $\pi_0\subset\pi\}$. We define the type of (π,π_0) as the type of the canonical map $(X,\emptyset)\to(\pi,\pi_0)$ sending an element of X to the surrounding block of π. Then

$t((k,0),\emptyset)=$ number of blocks in $\pi-\pi_0$ with k elements,
$t((k,0),J)=$ number of blocks in π_0 with k elements, and
$t((k,l),J)=0$ if $l>0$. Furthermore, $c(t)(J)=\#\pi_0$.

To every graph G with vertex set X we can associate a quotient object (π,π_0) of $(X,\emptyset)$ such that the blocks of π carry the connected components of G while the blocks of π_0 carry the connected components with property E. For any quotient object (π,π_0) of type t, the number of graphs giving rise to (π,π_0) in this way clearly is

$$f(t)=\Pi_{k=1}^{\infty}\ C(k,E)^{t((k,0),J)}(C(k)-C(k,E))^{t((k,0),\emptyset)}$$

where C(k) denotes the number of connected graphs with vertex set $\{1,..,k\}$. Since there are $G(\varepsilon((n,0),J);t,\varepsilon(c(t),J))$ quotient objects of $(X,\emptyset)$ of type t, we have

$$N(n,E;m) = \Sigma\, G(\varepsilon((n,0),J);t,\varepsilon(c(t),J))f(t) \quad ,$$

the summation being taken over all $t\in T$ such that $c(t)(J)=m$. With the multiplicative functions $f\in Mu(\mathbb{Z})$, $f(\varepsilon((k,0),\emptyset)) = C(k)-C(k,E)$, $f(\varepsilon((k,l),J)) = C(k,E)$ if $l=0$ and $= 0$ otherwise, and $g\in Mu(\mathbb{Z}[x])$, $g(\varepsilon((k,0),\emptyset)) = 0$, $g(\varepsilon((k,l),J)) = x^l$, it follows that

$$\Sigma_{m=0}^{n}\, N(n,E;m)x^m = fg(\varepsilon((n,0),J)) \quad \text{for } n>0 \quad ,$$

and hence

$$\Sigma_{n=1}^{\infty}\, (\Sigma_{m=0}^{n}\, N(n,E;m)x^m)\; z^n/n! = \rho(fg)(z(J))\,|_{z(\emptyset)=z,\; z(J)=0}$$

where $|_{..}$ means evaluation. But $\rho(g)(z(J)) = \exp(z(\emptyset)+xz(J)) -1$, $\rho(f)(z(\emptyset)) = \Sigma_{k=1}^{\infty}\, C(k)z(\emptyset)^k/k! - c_E(z(\emptyset))$ and $\rho(f)(z(J)) = c_E(z(\emptyset))$. Consequently,

$$\Sigma_{n=0}^{\infty}\, (\Sigma_{m=0}^{n}\, N(n,E;m)x^m)\; z^n/n! = \exp(\Sigma_{k=1}^{\infty}\, C(k)z^k/k! + (x-1)c_E(z)) \;.$$

By (2.28), Example a, $\exp(\Sigma_{k=1}^{\infty} C(k)z^k/k!) = \Sigma_{l=0}^{\infty}\, 2^{l(l-1)/2}z^l/l!$ which finishes the proof. □

§2. Invariant partitions of finite sets under group action

Let G be an abstract group. In this section we consider the category G-Sf of finite G-sets and G-homogeneous maps introduced in (2.4), Example e. In G-Sf a morphism $s:X\to Y$ is an epimorphism if and only if s is onto. Let M be the class of epimorphisms, and let ~ be the isomorphism relation on M.

Given a G-set X (We briefly speak of G-sets instead of finite G-sets), the group G operates on the lattice of partitions of the set X in a natural way: If $g\in G$ and $\pi=\{B_1,..,B_l\}$, then $g\pi=\{gB_1,..,gB_l\}$ where $gB_j=\{gx;x\in B_j\}$. A partition π of the set X is called invariant if all $g\in G$ fix π, i.e. permute the blocks of π. In particular, the partitions of X in G-Sf are all invariant because here G even fixes the blocks. Evidently, the set of invariant partitions of X is a sublattice of the partition lattice of the set X. We identify a quotient object $[r:X\to Y]$ of X with the invariant partition $\{r^{-1}(y);y\in Y\}$, and so the poset Qu(X) is the lattice of invariant partitions of X. When G is the trivial group and hence G-Sf the category of finite sets, Qu(X) is just the partition lattice of the set X.

Let $\underset{\sim}{C}$ be a system of representatives of the conjugacy classes of subgroups of finite index in G. Then

$$\underset{\sim}{P} = \{G/V;\ V\in\underset{\sim}{C}\}$$

constitutes a system of representatives of the isomorphism classes of indecomposable objects in G-Sf. The Krull-Schmidt type of a G-set X is $n\in \mathbb{N}_o(\underset{\sim}{P})$ where $n(G/V)$ equals the number of orbits of G in X whose stabilizers are conjugate to V. By (2.19) the order of the automorphism group of X is

$$a(n) = \Pi_{V\in\underset{\sim}{C}} [N_G(V):V]^{n(G/V)} n(G/V)!$$

where $[N_G(V):V]$ denotes the index of V in its normalizer.
Our next aim is to count invariant partitions of G-sets. For $V,W\in\underset{\sim}{C}$, let

$$(V,W) = \#\{gV\in G/V;\ g^{-1}Wg\subset V\} \quad .$$

In particular, $(V,V)=[N_G(V):V]$. Define a partial order on $\underset{\sim}{C}$ by

$W\leq V$ if and only if W is conjugate to a subgroup of V .

Then $(V,W)=0$ unless $W\leq V$. The numbers (V,W) are the so-called marks of Burnside ([Bur,11]) which nowadays appear in the Pólya-Redfield enumeration theory (see [KT,82]).

(4.6) PROPOSITION: For $n\in \mathbb{N}_o(\underset{\sim}{P})$ let $B(n)$ denote the number of invariant partitions of a G-set of KS-type n (If G is the trivial group, the $B(n)$ are the Bell numbers). Then

$$\Sigma_n\ B(n)\ z^n/n! = \exp(\Sigma_V\ (V,V)^{-1}\{\exp(\Sigma_{W\leq V}(V,W)z(G/W)) -1\})$$

in the power series algebra $\mathbb{Q}[[z(G/V);V\in\underset{\sim}{C}]]$.

By differentiating, a recursion formula for the numbers $B(n)$ can be derived which generalizes the customary recursion formula for the Bell numbers.

(4.7) COROLLARY: $B(0)= 1$ and

$$B(n+\varepsilon(G/V))= \Sigma_{1\leq n} \binom{n}{1} B(n-1)\{\Sigma_{Z\geq V} \frac{(Z,V)}{(Z,Z)} \Pi_{W\in\underset{\sim}{C}}(Z,W)^{1(G/W)}\}$$

where $1\leq n$ means $1(G/W)\leq n(G/W)$ for all $W\in\underset{\sim}{C}$, and where $\binom{n}{1}= \Pi_{W\in\underset{\sim}{C}} \binom{n(G/W)}{1(G/W)}$ is a product of binomial coefficients. □

Proof of Proposition (4.6): In order to apply (3.39) we count, for non-zero $n\in \mathbb{N}_o(\underset{\sim}{P})$ and $V\in\underset{\sim}{C}$, the G-homogeneous maps $s:X(n)\to G/V$ where $X(n)= \oplus_V n(G/V)G/V$. Denoting the set of G-homogeneous maps from X to Y by $Hom(X,Y)$, we have $\#Hom(X(n),G/V)= \Pi_W\ \#Hom(G/W,G/V)^{n(G/W)}$. But a G-homogeneous map $s:G/W\to G/V$ is uniquely determined by the coset $s(1W)=gV$ which satisfies $g^{-1}Wg\subset V$, and, conversely, each coset gV such that $g^{-1}Wg\subset V$ gives rise to a G-homogeneous map $G/W\to G/V$, $hW\mapsto hgW$. Thus $\#Hom(G/W,G/V)=(V,W)$ and $\#M(n,\varepsilon(G/V))= \#Hom(X(n),G/V)= \Pi_W(V,W)^{n(G/W)}$. Substituting $iq(n)= \Sigma_V\ (V,V)^{-1}\ \Pi_W(V,W)^{n(G/W)}$ into (3.39), we obtain the stated formula. □

We now determine the affine monoid of multiplicative functions, but in the special case that G is commutative. For an arbitrary group G, it seems difficult to describe the indecomposable types and to calculate the elementary section coefficients in an explicit way.
Henceforth G is an abelian group, with addition + and unit 0. Then $\underset{\sim}{C}$ is the set of subgroups of finite index in G ordered by inclusion, and G acts on G-sets by automorphisms. Any indecomposable epimorphism in G-Sf is isomorphic to exactly one of the morphisms
$\oplus_W n(G/W)G/W \to G/V$, $g+W \mapsto g+V$ on each orbit, where n is non-zero and $n(G/W)=0$ for all $W \in \underset{\sim}{C}$ which are not contained in V.
Thus we are in the situation of (3.44) with

$$S=\{(n,G/V); n \in \mathbb{N}_o(\underset{\sim}{P}), V \in \underset{\sim}{C} \text{ such that } n \neq 0 \text{ and } n(G/W)=0 \text{ unless } W \subset V\}$$
$$\text{and} \qquad T=\mathbb{N}_o(S) \quad .$$

To compute the elementary section coefficients (3.45), let $X(n)=\oplus_W n(G/W)G/W$ and consider $s:X(n)\to G/V$, $g+W\mapsto g+V$ on each orbit. Then the invariant partition $\sigma=\{s^{-1}(g+V); g\in G\}$ of $X(n)$ contains $s^{-1}(0+V)=\oplus_W n(G/W)V/W$. By (1.40),
$G(\varepsilon(n,G/V);t,\varepsilon(c(t),G/V))=\#\{\pi \in Qu(X(n)); \pi \leq \sigma \text{ and } \overline{\pi}=t\}$. Here $\overline{\pi}$ is defined as the type of the canonical map $X(n)\to\pi$ sending an $x\in X(n)$ to the surrounding block of π. But any G-invariant partition π of the G-set $X(n)=\oplus_W n(G/W)G/W$ which is less or equal σ induces a V-invariant partition of the V-set $X(n)^*=\oplus_W n(G/W)V/W$, namely $\pi^*=\{B\in\pi; B\subset X(n)^*\}$. Conversely, every V-invariant partition π^* of $X(n)^*$ can be extended to the G-invariant partition $\pi=\{\{g+x; x\in B\}; g\in B, B\in\pi^*\}$ of $X(n)$. Since these constructions are mutually inverse, we have a one-to-one correspondence between G-invariant partitions of $X(n)$ which are less or equal σ, and V-invariant partitions of $X(n)^*$. Hence we consider also the category V-Sf of finite V-sets where $\underset{\sim}{P}^*$, S^* and T^* are defined as $\underset{\sim}{P}$, S and T in G-Sf. If π has the type $t\in T$, then for any $(m,G/W)\in S$ there are exactly $t(m,G/W)$ orbits b of G on π whose stabilizer is W and where $\cup_{B\in b}B \subset X(n)$ has the KS-type m. Thus the type of $\pi^*=\{B\in\pi; B\subset X(n)^*\}$ is $t^*\in T^*$ where, for $(m^*,V/W)\in S^*$, $t^*(m^*,V/W)=t(m,G/W)$, $m\in\mathbb{N}_o(\underset{\sim}{P})$ being given by $m(G/W)=m^*(V/W)$ if $W\subset V$ and $m(G/W)=0$ otherwise. It follows that $G(\varepsilon(n,G/V);t,\varepsilon(c(t),G/V))=\#\{\pi^*\in Qu(X(n)^*); \overline{\pi^*}=t^*\}$. Now we can apply (3.52) in V-Sf and obtain

$$G(\varepsilon(n,G/V);t,\varepsilon(c(t),G/W))=\frac{n!}{t!}\prod_{(m,G/W)\in S}([V:W]^{|m|-1}/m!)^{t(m,G/W)}$$

where $[V:W]$ denotes the index of W in V. Using the multiplicative function $\varphi:T\to E(\mathbb{Q})$, $\varphi(\varepsilon(n,G/V))=\prod_{W\subset V}[V:W]^{n(G/W)}n(G/W)!$ for $(n,G/V)\in S$, this can be written as $c(t)!\varphi(\varepsilon(n,G/V))/t!\varphi(t)\varphi(\varepsilon(c(t),G/V))$.

Hence, by Proposition (3.48), we have the following result.

(4.8) THEOREM: The affine monoid of multiplicative functions on T has the faithful power series representation

$$\rho_R: \mathrm{Mu}(R) \to \mathrm{End}(R[[z(G/V);V\in \underset{\sim}{C}]]) \quad , \ R\in \mathrm{Al}_{\mathbb{Q}} \ ,$$
$$f \to [z(G/V) \to \Sigma_n \ f(\varepsilon(n,G/V))z^n/a_V(n)] \quad .$$

Here $a_V(n) = \Pi_{W\subset V}[V:W]^{n(G/W)}n(G/W)!$ is the order of the automorphism group of $\oplus_W n(G/W)V/W$ in V-Sf, and [V:W] denotes the index of W in V. □

Under the representation ρ the zeta function $\zeta\in\mathrm{Mu}''(\mathbb{Z})$ acts by the transformation $[z(G/V) \to \exp(\Sigma_{W\subset V}[V:W]^{-1}z(G/W))-1]$. To compute the Möbius function $\mu=\zeta^{-1}$, we have to invert $\rho(\zeta)$, i.e. to solve the equations $\exp(\Sigma_{W\subset V}[V:W]^{-1}\rho(\mu)(z(G/W)))-1 = z(G/V)$, $V\in\underset{\sim}{C}$, or

$$(*) \qquad \Sigma_{W\subset V}[V:W]^{-1}\rho(\mu)(z(G/W)) = \log(1+z(G/V)) \ , \ V\in\underset{\sim}{C} \ .$$

The set $\underset{\sim}{C}$ of all subgroups of finite index in G becomes, when ordered by inclusion, a locally finite join-semilattice with the greatest element G. Applying Möbius inversion on $\underset{\sim}{C}$ to the power series identity (*) for every coefficient, it follows that

$$\rho(\mu) = [z(G/V) \to \Sigma_{W\subset V} \ \mu_{\underset{\sim}{C}}(W,V)[V:W]^{-1}\log(1+z(G/W))]$$

where $\mu_{\underset{\sim}{C}}$ denotes the Möbius function on $\underset{\sim}{C}$.

Expanding log into a power series, we can read off the values of μ on the indecomposable types:

$$\mu(\varepsilon(n,G/V)) = (-1)^{l-1}(l-1)! \ \mu_{\underset{\sim}{C}}(W,V)[V:W]^{l-1}$$

(4.9) $\qquad$ if $n=l\varepsilon(G/W)$ where $l\in\mathbb{N}$ and $W\subset V$,

and $\mu(\varepsilon(n,G/V)) = 0$ else .

Now let X be a G-set of KS-type n. The lattice Qu(X) of invariant partitions of X has the least element $\hat{0}=\{\{x\};x\in X\}$ and the greatest element $\hat{1}=\{X\}$. Since the canonical map $\{\{x\};x\in X\}\to\{X\}$ has type $\varepsilon(n,G/G)$, Theorem (1.44) implies that the Möbius function of Qu(X) takes the value

$$\mu_{Qu(X)}(\hat{0},\hat{1}) = (-1)^{l-1}(l-1)! \ \mu_{\underset{\sim}{C}}(W,G)[G:W]^{l-1}$$

(4.10) $\qquad$ if $n=l\varepsilon(G/W)$ where $l\in\mathbb{N}$,

and $\mu_{Qu(X)}(\hat{0},\hat{1}) = 0$ else .

Finally, we consider the special case when <u>G= $\mathbb{Z}$, the infinite cyclic group</u>. A $\mathbb{Z}$-set X can be viewed as a finite set X with a permutation σ

of X, the connection being given by $gx=\sigma^g(x)$ for $g\in\mathbb{Z}$ and $x\in X$. Thus a partition π of the set X is invariant if and only if σ permutes the blocks of π. The lattice of invariant partitions of X has been studied recently by P.Hanlon ([Han,81]).
As $\underset{\sim}{C}=\{\mathbb{Z}i;i\in\mathbb{N}\}$, a system of representatives of the isomorphism classes of indecomposable objects in $\mathbb{Z}$-Sf is

$$\underset{\sim}{P}=\{\mathbb{Z}/\mathbb{Z}i;i\in\mathbb{N}\} \quad .$$

The Krull-Schmidt type of a $\mathbb{Z}$-set X is $n\in\mathbb{N}_o(\underset{\sim}{P})$ where

$$n(\mathbb{Z}/\mathbb{Z}i) = \text{number of cycles of } \sigma \text{ of length } i \quad , \quad i\in\mathbb{N} \quad .$$

To simplify the notation, we use instead of KS-types $n\in\mathbb{N}_o(\underset{\sim}{P})$ cycle types $\alpha\in\mathbb{N}_o(\mathbb{N})$ written as $\alpha=(1^{\alpha(1)}2^{\alpha(2)}\ldots)$. When specializing the above results, observe that the poset $\underset{\sim}{C}$ is anti-isomorphic to the set $\mathbb{N}$ of positive integers ordered by divisibility.
For instance, if X is a finite set and if σ is a permutation of X of cycle type α, then $B(\alpha)$ counts the invariant partitions of X. The recursion formula (4.7) gives

$$B(0)=1 \quad \text{and} \quad B(\alpha+\varepsilon(i))=\Sigma_\beta\binom{\alpha}{\beta}B(\alpha-\beta)\{\Sigma_d\, d^{|\beta|}\} \quad .$$

Here the outer sum runs over all $\beta\in\mathbb{N}_o(\mathbb{N})$ such that $\beta(j)\leq\alpha(j)$ for all j, and the inner sum over all (positive) divisors d of i which divide every j with $\beta(j)\neq 0$. As usual, $|\beta|=\Sigma_{i=1}^{\infty}\beta(i)$.
Expressing the Möbius function on $\underset{\sim}{C}$ by the number-theoretic Möbius function $\mu:\mathbb{N}\to\mathbb{Z}$, we infer from (4.10) that

$$\mu_{Qu(X)}(\hat{0},\hat{1}) = (-1)^{l-1}(l-1)!\ \mu(i)i^{l-1} \quad \text{if } \alpha=l\varepsilon(i) \text{ where } l\in\mathbb{N} ,$$
$$\text{and} \quad = 0 \quad \text{otherwise} .$$

This result has been derived combinatorically by P.Hanlon in [Han,81], and has been applied by R.P.Stanley in [St,82] to special linear representations of the symmetric group on the homology groups of the partition lattices.
We conclude with a numerical example.
Let X={1,2,3,4} and let $\sigma=(12)(34)$ which has cycle type $\alpha=(1^0 2^2 3^0 4^0\ldots)$.

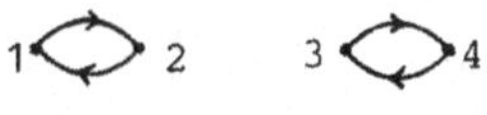

Here the invariant partitions are
$\hat{1}=\{\{1,2,3,4\}\}$,
$\{\{1,2\},\{3,4\}\}$,
$\{\{1,3\},\{2,4\}\}$, $\{\{1,4\},\{2,3\}\}$,
$\{\{1\},\{2\},\{3,4\}\}$, $\{\{1,2\},\{3\},\{4\}\}$,
$\hat{0}=\{\{1\},\{2\},\{3\},\{4\}\}$,

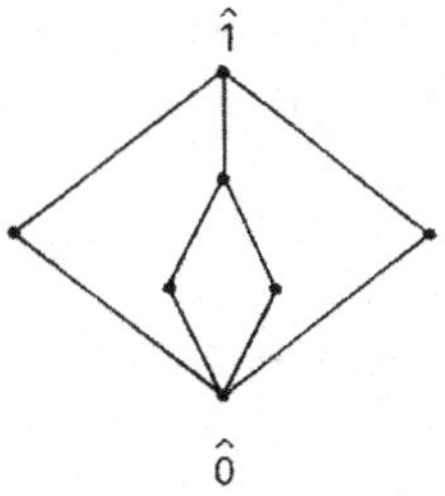

while $B(\alpha)=2B(\varepsilon(2))+3B(0)=2\cdot 2+3\cdot 1=7$
and $\mu_{Qu(X)}(\hat{0},\hat{1})=(-1)^1 1!\ \mu(2)2^1=2$.

§3. Rooted forests and Butcher series

By a (finite undirected simple) graph we understand a pair (V,E) where V (the vertex set) is a finite set and E (the edge set) is a set of 2-element subsets of V. A rooted tree is a connected graph without cycles with a distinguished vertex called the root.

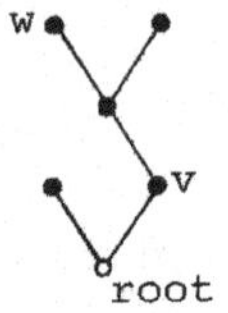

A rooted forest is a graph whose connected components are rooted trees. Its vertex set is partially ordered by $v \leq w$ if and only if the path from w to the root visits v. Given a rooted forest X, a subforest of X is a rooted forest which is a subgraph of X and whose roots are also roots of X. Finally, a morphism from one rooted forest to another is a map between the vertex sets preserving roots, neighbors and the order relation.

Let $\underline{Fr}$ denote the category of rooted forests. The initial object 0 of $\underline{Fr}$ is the empty forest, and the direct sum in $\underline{Fr}$ is formed by taking the disjoint union of the vertex sets and the edge sets. Thus the indecomposable objects in $\underline{Fr}$ are the rooted trees. Since the monomorphisms in $\underline{Fr}$ are one-to-one, we can identify a subobject $[s:X\to Y]$ of a rooted forest Y with the subforest s(X) whose vertex set is $\{s(v); v$ a vertex of $X\}$, whose edge set is $\{\{s(e_1),s(e_2)\};\{e_1,e_2\}$ an edge of $X\}$ and whose roots are the s(r), r a root of X. Then the unique Krull-Schmidt partition of a rooted forest is the set of its connected components.

If Y is a rooted forest and if Z is a subforest of Y, a new rooted forest Y-Z is obtained by removing from Y all vertices of Z and all adjacent edges of Y, and by choosing as roots of Y-Z the minimal vertices left over.

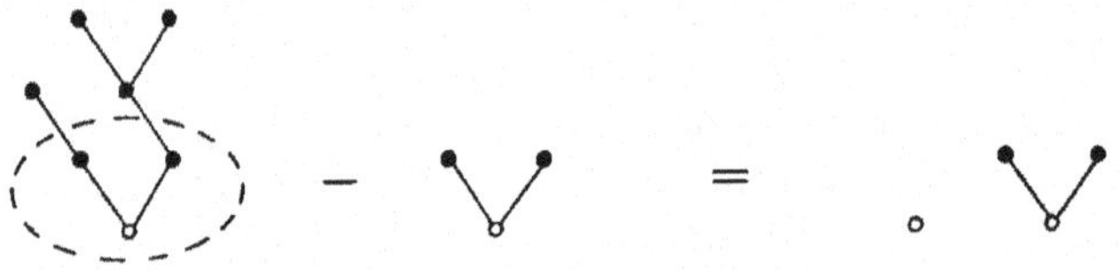

On the class M of all monomorphisms in $\underline{Fr}$, we consider the following equivalence relation: $(s_1:X_1\to Y_1)\sim(s_2:X_2\to Y_2)$ if and only if $Y_1-s_1(X_1)$ and $Y_2-s_2(X_2)$ are isomorphic. Let

$\underset{\sim}{P}$ = { 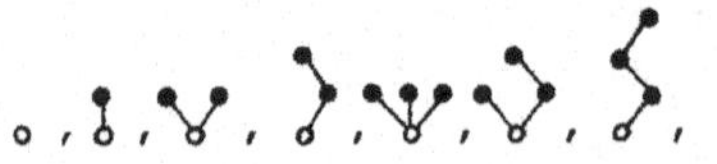 , .. }

be a system of representatives of the isomorphism classes of rooted trees. Then the monoid of types $T=M/\sim$ is isomorphic to the monoid $\mathbb{N}_o(\underset{\sim}{P})$ of KS-types by $\overline{s:X\to Y} \mapsto$ KS-type of Y-s(X) , $\overline{0\to\oplus_P n(P)P} \leftarrow n$, and we identify types with KS-types. Hence

$$T = \mathbb{N}_o(\underset{\sim}{P}) \quad\text{and}\quad U = \{\varepsilon(P); P\in\underset{\sim}{P}\} \quad.$$

If n,n_1,n_2 are types and if Y is a rooted forest of KS-type n, then by (1.33) the section coefficient $G(n;n_1,n_2)$ counts the subforests of Y of KS-type n_1 after whose removal a rooted forest of KS-type n_2 is left. E.g. $G(\vee;\, \overset{\bullet}{\circ}\,,\,o\,)=2$ whereas $G(\vee;\,o\,,\,\overset{\bullet}{\circ}\,)=0$
(To simplify the notation, we write P instead of $\varepsilon(P)$ for $P\in\underset{\sim}{P}$).

According to (3.29), the incidence bialgebra $\mathbb{Z}(T;G)$ is a topological Hopf algebra, and the affine monoid Mu of multiplicative functions on T is a unipotent group defined over $\mathbb{Z}$. For $R\in Al_{\mathbb{Z}}$, the convolution product of two multiplicative funciions $f,g:T\to R$ is given by

$$fg(P) = \Sigma_{Q,n} G(P;Q,n) f(Q) g(n) + g(P) \quad , \quad P\in\underset{\sim}{P} \quad ,$$

where the sum runs over all $Q\in\underset{\sim}{P}$ and $n\in\mathbb{N}_o(\underset{\sim}{P})$. For instance,

$$fg(\vee) = f(\vee) + f(\overset{\bullet}{\curlyvee})g(_o) + f(\vee)g(_o) + f(\overset{\bullet}{\circ})g(\overset{\bullet}{\circ}) + f(\overset{\bullet}{\circ})g(_o)^2 +$$
$$+ f(_o)g(_o)g(\overset{\bullet}{\circ}) + g(\vee) \quad .$$

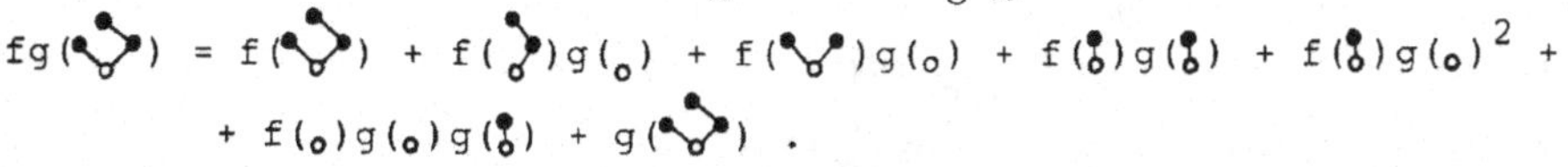

To determine the dimension of a type $n\in\mathbb{N}_o(\underset{\sim}{P})$, we choose the representative $s:0\to\oplus_P n(P)P$. By (1.32) the dimension of s equals the maximal length of a chain from 0 to $\oplus_P n(P)P$ in the poset of subforests of $\oplus_P n(P)P$, which implies that

$$\dim(n) = \text{number of vertices of } \oplus_P n(P)P \quad .$$

By Proposition (3.26) the group Mu is filtered by the normal subgroups $Mu(d)(R) = \{f\in Mu(R); f(P)=0 \text{ for all } P\in\underset{\sim}{P} \text{ with less than d vertices}\}$, $d\in\mathbb{N}$, and is the inverse limit of the factor groups $Mu/Mu(d)$. Here each $Mu/Mu(d)$ is a unipotent algebraic group because for $f,g\in Mu(R)$, $f\equiv g$ mod $Mu(d)$ if and only if $f(P)=g(P)$ for all $P\in\underset{\sim}{P}$ with less than d vertices. The Lie algebra of Mu has the topological basis $(e(P); P\in\underset{\sim}{P})$ with the bracket $\quad [e(P_1),e(P_2)] = \Sigma_P\{G(P;P_1,P_2)-G(P;P_2,P_1)\}e(P) \quad .$
In this sum only rooted trees occur which are patched together from P_1 and P_2, e.g. $[e(\vee),e(\overset{\bullet}{\circ})] = e(\overset{\bullet}{\curlyvee}) + e(\curlyvee) - e(\overset{\bullet}{\curlyvee}) - e(\curlyvee)$.

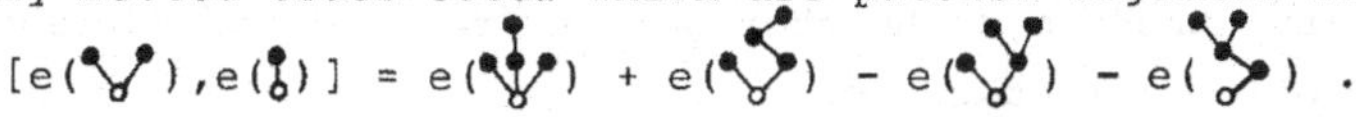

The group Mu has an interesting application to numerical mathematics, in particular to the numerical integration of ordinary differential equations ([But,72] and [HW,74]). In the sequel we give a brief introduction to the theory of Butcher series.
Let Y be an open subset of $\mathbb{R}^m$, let $F:Y\to\mathbb{R}^m$ be infinitely differentiable, let $x_o\in\mathbb{R}$ and $y_o\in Y$, and consider the initial value problem

$$(*) \quad y' = F(y) \quad , \quad y(x_o) = y_o \quad .$$

For each $P\in\underset{\sim}{P}$, the "elementary differential" $DF(P):Y\to\mathbb{R}^m$ is defined recursively by

$DF(\circ)(y) = F(y)$ and

$DF(P)(y) = F^{(r)}(y)(DF(P_1)(y),..,DF(P_r)(y))$

where $F^{(r)}(y)$ is the r-th total derivative of F in y and where $P_1,..,P_r \in \underset{\sim}{P}$ are the branches of P, up to isomorphism.

P = 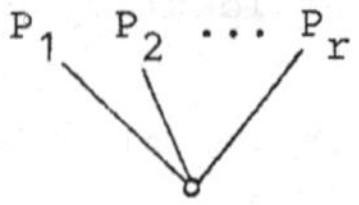

For $n \in \mathbb{N}_o(\underset{\sim}{P})$, let mon(n) be the number of order-preserving bijections from the vertex set of $\oplus_P n(P)P$ to the totally ordered set $\{1,..,\dim(n)\}$, or, in other words, the number of monotonous labelings of the rooted forest $\oplus_P n(P)P$. Further, let a(n) denote the order of the automorphism group of $\oplus_P n(P)P$.

By differentiating in (*), any derivative of the solution y can be expressed in terms of F. J.C.Butcher gives the explicit formula

$$y^{(d)}(x_o) = \Sigma \frac{mon(P)}{a(P)} DF(P)(y_o) \quad ,$$

the summation being taken over all $P \in \underset{\sim}{P}$ with dim(P)=d.

E.g. $y''(x_o) = \frac{1}{1}DF($$)(y_o) = F'(y_o)(F(y_o))$ and

$y'''(x_o) = \frac{2}{2}DF($$)(y_o) + \frac{1}{1}DF($$)(y_o) =$

$= F''(y_o)(F(y_o),F(y_o)) + F'(y_o)(F'(y_o)(F'(y_o)))$.

Consequently, the Taylor series of y around x_o is

$$\Sigma_{d=0}^{\infty} y^{(d)}(x_o)h^d/d! = y_o + \Sigma_{P\in\underset{\sim}{P}} \frac{mon(P)}{\dim(P)!} DF(P)(y_o)\, h^{\dim(P)}/a(P) \quad .$$

Formal series

$$y_o + \Sigma_{P\in\underset{\sim}{P}} \lambda(P)DF(P)(y_o)\, h^{\dim(P)}/a(P)$$

where the $\lambda(P)$ are arbitrary real numbers, are called <u>Butcher series</u> by E.Hairer and G.Wanner, and used to study special integration methods for (*), in particular Runge-Kutta methods. The composition of these methods is described by the formal composition of the corresponding Butcher series. For instance, if $\lambda(\circ)=1$ and $\lambda(P)=0$ else, we obtain Euler's method which approximates $y(x_o+h)$ by $y_o+hF(y_o)$.

The central theorem on Butcher series states that, for every initial value problem (*), the map

$$(4.11) \qquad f \to y_o + \Sigma_{P\in\underset{\sim}{P}} f(P)DF(P)(y_o)\, h^{\dim(P)}/a(P) \quad , \quad f \in Mu(\mathbb{R}) \quad ,$$

is a group homomorphism. Thus it is possible to study the composition of methods inside $Mu(\mathbb{R})$. Furthermore, using the filtration

$$Mu(\mathbb{R}) = Mu(1)(\mathbb{R}) \supset Mu(2)(\mathbb{R}) \supset Mu(3)(\mathbb{R}) \supset \dots \quad ,$$

the concept of order can be defined for these integration methods. For details we refer the reader to [But,72] and [HW,74].

§4. Matroids and graded Hopf algebras

In this section we sketch how graded (topological) Hopf algebras arise from special classes of matroids, following an idea of S.A.Joni and G.C.Rota in [JR,79]. In the sequel we use some elementary concepts of matroid theory which can be found in [CR,71],[Tu,70] or [We,76]. Recall that a matroid is a finite set X with a closure relation $\bar{\ }$ that enjoys the MacLane-Steinitz exchange property. A morphism from one matroid X to another matroid Y is a map $s:X\to Y$ such that $s(\bar{A})\subset\overline{s(A)}$ for every subset A of X. The category of matroids Mat is a Krull-Schmidt category with unique KS-partitions.

Now, let K be a class of matroids having the following properties:

(1) K is closed under isomorphism, direct sum and decomposition into direct summands.

(2) Reduction or contraction of a matroid in K by closed subsets again yields a matroid in K.

Examples are the classes of free matroids, graphical matroids, unimodular matroids and others (see [JR,79]). We denote the full subcategory of Mat with objects in K by $\underline{K}$. Let M be the class of all monomorphisms $s:X\to Y$ in $\underline{K}$ which have a closed image and map X isomorphically to the reduction $Y|s(X)$. For any matroid Y in K, a subobject $[r:X\to Y]$ of Y in M can be identified with the closed subset $r(X)$ of Y. So $\mathrm{Sub}_M(Y)$ is the set of closed subsets of Y ordered by inclusion which is known to be a geometric lattice.

On M we define an equivalence relation $\sim$ by $(s_1:X_1\to Y_1)\sim(s_2:X_2\to Y_2)$ if and only if the combinatorial geometries of the contractions $Y_1-s_1(X_1)$ and $Y_2-s_2(X_2)$ are isomorphic.

Then $\underline{K},M,\sim$ is a sheaflike categorical structure where the types are in one-to-one correspondence to the isomorphism types of the combinatorial geometries associated to the matroids in K. By (3.29) the affine monoid Mu of multiplicative functions is a unipotent group defined over the integers, and the incidence bialgebra $H=\mathbb{Z}(T;G)$ is a topological Hopf algebra which is graded by the sequence of closed submodules

$$V(d) = \{f\in\mathbb{Z}^T; f(t)=0 \text{ unless } \dim(t)=d\} \quad , \quad d\in\mathbb{N} \quad .$$

Here the dimension of a type $t\in T$ equals the rank of a combinatorial geometry of the corresponding isomorphism type. The contravariant Hopf algebra of Mu which is also graded, is called a "hereditary bialgebra" by Joni and Rota, and proposed as an appropriate frame for a "general umbral calculus" (compare [NS,80]).

§5. Partial partitions of finite sets

Let N be a subset of $\mathbb{N}=\{1,2,3,..\}$. An N-partition of a finite set X is a partition of X such that the cardinality of each block belongs to N. We consider the following category $\underline{K}$: The objects in $\underline{K}$ are pairs (X,π) where X is a finite set and π is an N-partition of X. A morphism from (X,π) to (Y,σ) is a function $s:X\to Y$ mapping each block of π into a block of σ.

In $\underline{K}$ the initial object $0=(\emptyset,\emptyset)$ is the empty set with the empty partition, and the direct sum $(X,\pi)\oplus(Y,\sigma)=(X\dot{\cup}Y,\pi\dot{\cup}\sigma)$ is formed by disjoint union. An object (Z,τ) is indecomposable if and only if τ has only one block. Thus a system of representatives of the indecomposable objects in $\underline{K}$ modulo isomorphism is given by

$$\underset{\sim}{P} = \{P_m;m\in N\} \qquad \text{where} \qquad P_m=(\{1,..,m\},\{\{1,..,m\}\}) \quad .$$

The Krull-Schmidt type $n\in \mathbb{N}_o(\underset{\sim}{P})$ of an object (X,π) tells, for every $m\in N$, how many blocks of π have m elements. For the sake of simplicity, we use instead of n the "type" $\alpha\in \mathbb{N}_o(N)$ of the partition π which is written as $\alpha=(j^{\alpha(j)};j\in N)$. For instance, if $N=\{1,3,5,7,..\}$, then $(X,\pi)=(\{1,2,3,4,5,6\},\{\{1\},\{2\},\{3\},\{4,5,6\}\})$ has the KS-type $\alpha=(1^3 3^1 5^0 7^0..)$. Note that, for an object (X,π) of KS-type α, $\omega(\alpha)=\Sigma_{j\in N}\, j\alpha(j)$ is the cardinality of X whereas $|\alpha|=\Sigma_{j\in N}\,\alpha(j)$ is the number of blocks of π. By (2.17) the automorphism group of (X,π) is isomorphic to the direct product of wreath products $\Pi_{j\in N}\, S_j \text{ wr } S_{\alpha(j)}$ and has the order

$$a(\alpha) = \Pi_{j\in N}\, (j!)^{\alpha(j)}\alpha(j)! \quad .$$

Let M be the class of monomorphisms in $\underline{K}$. In $\underline{K}$ a morphism $s:(X,\pi)\to(Y,\sigma)$ is a monomorphism if and only if $s:X\to Y$ is one-to-one. By a partial N-partition of Y, we understand an N-partition of a subset of Y (In the special case $N=\mathbb{N}$, we simply speak of a partial partition of Y). Given two partial N-partitions π,τ of Y, π is called finer than τ (denoted by $\pi\leq\tau$) if each block of π is contained in a block of τ. We identify a subobject $[r:(X,\pi)\to(Y,\tau)]$ of (Y,σ) with the partial N-partition $\{r(B);B\in\pi\}$ of Y. Then $Sub(Y,\sigma)$ is the set of all partial N-partitions of Y finer than σ, ordered by refinement. The least element of $Sub(Y,\sigma)$ is $\emptyset$, the greatest element is σ.

Since $Sub(Y,\sigma)$ is order-isomorphic to the product $\Pi_{B\in\sigma}\, Sub(B,\{B\})$, it suffices to consider the posets $Sub(P_m)$ consisting of all partial N-partitions of $\{1,..,m\}$, $m\in N$. Then the set Π_m^N of all N-partitions of $\{1,..,m\}$ is a subset of $Sub(P_m)$ with the property that every interval of Π_m^N also is an interval of $Sub(P_m)$. If $N=\mathbb{N}$, the partition lattice Π_m is a sublattice of the lattice of partial partitions of $\{1,..,m\}$.

Finally, let $\sim$ be the isomorphism relation on M. The isomorphism type of an indecomposable monomorphism $s:(X,\pi)\to(Y,\{Y\})$ is uniquely determined by the KS-type α of (X,π) and the cardinality m of Y. Since s is one-to-one, $\#X=\omega(\alpha)\leq m=\#Y$. Thus we are in the situation of (3.44) with

$$S=\{(\alpha,P_m);\alpha\in \mathbb{N}_o(N), m\in N \text{ such that } \omega(\alpha)\leq m\} \quad \text{and} \quad T=\mathbb{N}_o(S) \quad .$$

To compute the elementary section coefficients (3.45), we choose a set Y with m elements and an N-partition π of a subset X such that (X,π) has KS-type α. Then the injection $s:(X,\pi)\to(Y,\{Y\})$ is a representative of $\varepsilon(\alpha,P_m)$, and, by (1.33), $G(\varepsilon(\alpha,P_m);t,\varepsilon(c(t),P_m))=\#A(Y,\pi,t)$ where $A(Y,\pi,t)$ is the set of all N-partitions τ of some subset Z which are coarser than π and such that the injection $(X,\pi)\to(Z,\tau)$ has type t. Now, let G be the group of all permutations of Y mapping X to itself as an automorphism of (X,π), and consider the operation of G on $A(Y,\pi,t)$ given by $g\tau=\{g(B);B\in\tau\}$ for $g\in G$ and $\tau\in A(Y,\pi,t)$. Evidently, G is isomorphic to the direct product of the automorphism group of (X,π) and the symmetric group on Y-X, and has the order $a(\alpha)(m-\omega(\alpha))!$. Moreover, it is not difficult to see that G acts transitively on $A(Y,\pi,t)$ and that the stabilizer of any $\tau\in A(Y,\pi,t)$ is isomorphic to the direct product of wreath products $(\Pi_{(\beta,n)\in S}\, G_{\beta,n} \text{ wr } S_{t(\beta,n)})\times S_{m-\omega(c(t))}$

where $G_{\beta,n}=\mathrm{Aut}(\oplus_{j\in N}\beta(j)P_j)\times S_{n-\omega(\beta)}$, and has the order

$o(m;t)=(\Pi_{(\beta,n)\in S}(a(\beta)(n-\omega(\beta))!)^{t(\beta,n)}t(\beta,n)!)(m-\omega(c(t)))!$.

We conclude that

$$G(\varepsilon(\alpha,P_m);t,\varepsilon(c(t),P_m))=\frac{a(\alpha)(m-\omega(\alpha))!}{o(m;t)} \quad .$$

Using the multiplicative function $\varphi:T\to E(\mathbb{Q})$, $\varphi(\varepsilon(\alpha,P_m))=\alpha!(m-\omega(\alpha))!$ for $(\alpha,P_m)\in S$, this can be written as

$c(t)!\varphi(\varepsilon(\alpha,P_m))/t!\varphi(t)\varphi(\varepsilon(c(t),P_m))$.

By Proposition (3.46) we get the following result.

(4.12) THEOREM: The affine monoid of multiplicative functions on T has the faithful representation

$$\rho_R: \mathrm{Mu}(R) \to \mathrm{End}(R[z_m;m\in N]) \quad , \quad R\in\mathrm{Al}_{\mathbb{Q}} \; ,$$
$$f \to [z_m \to \Sigma_\alpha\, f(\varepsilon(\alpha,P_m))z^\alpha/\alpha!(m-\omega(\alpha))!]$$

by endomorphisms of the polynomial algebra $\mathbb{Q}[z_m;m\in N]$. □

From this a faithful power series representation of Mu can be derived. For any $f\in\mathrm{Mu}(R)$, $\rho_R(f)$ extends to an endomorphism $\rho_R'(f)$ of $R[w,z_m;m\in N]$ by $\rho_R'(f)(w)=w$, w being a new indeterminate. Conjugating with the automorphism Φ of the quotient field $\mathbb{Q}(w,z_m;m\in N)$ given by $\Phi(w)=w$ and

$\Phi(z_m) = z_m/w^m$, we obtain the endomorphism $\rho''_R(f)$ of $R[w,z_m;m\in N]$ which fixes w and sends z_m to $\Sigma_\alpha\ f(\varepsilon(\alpha,P_m))\ z^\alpha/\alpha!\ w^{m-\omega(\alpha)}/(m-\omega(\alpha))!$. But this defines an endomorphism $\hat{\rho}_R(f)$ of the power series algebra $R[[w,z_m;m\in N]]$.

(4.13) COROLLARY: The affine monoid of multiplicative functions on T admits the faithful power series representation

$$\hat{\rho}_R\colon Mu(R) \to End(R[[w,z_m;m\in N]]) \quad , \quad R\in Al_{\mathbb{Q}} \ ,$$
$$f \mapsto \begin{bmatrix} w \to w \\ z_m \to \Sigma_{n=0}^{m}(\Sigma_{\alpha,\omega(\alpha)=m-n}\ f(\varepsilon(\alpha,P_m))z^\alpha/\alpha!)\ w^n/n! \end{bmatrix} \ . \ \square$$

Under this representation the zeta function $\zeta\in Mu''(\mathbb{Z})$ acts by the transformation

$$\hat{\rho}(\zeta) = \begin{bmatrix} w \to w \\ z_m \to \frac{1}{m!}(\partial^m/\partial x^m)|_{x=0}\ \exp(wx+\Sigma_{j\in N}z_jx^j) \end{bmatrix}$$

where x is a new variable and $(\partial^m/\partial x^m)|_{x=0}$ means differentiation followed by setting x=0. Consequently, the values of the Möbius function $\mu=\zeta^{-1}$ can be calculated by solving the equations

$$(4.14) \qquad \frac{1}{m!}(\partial^m/\partial x^m)|_{x=0}\ \exp(wx+\Sigma_{j\in N}\hat{\rho}(\mu)(z_j)x^j) = z_m \quad , \quad m\in N \ ,$$

for the power series $\hat{\rho}(\mu)(z_m)$, $m\in N$, in $\mathbb{Q}[[w,z_m;m\in N]]$. We will accomplish this in special cases.

Case 1: $N=\mathbb{N}=\{1,2,3,..\}$.

From (4.14) it follows that $\exp(wx+\Sigma_{j=1}^\infty\hat{\rho}(\mu)(z_j)x^j) = 1+\Sigma_{m=1}^\infty z_mx^m$ and $\Sigma_{j=1}^\infty\hat{\rho}(\mu)(z_j)x^j = \log(1+\Sigma_{m=1}^\infty z_mx^m) - wx$. Comparing coefficients, we find that

$$\hat{\rho}(\mu) = \begin{bmatrix} w \to w \\ z_1 \to z_1 - w \\ z_m \to \Sigma_{\alpha,\omega(\alpha)=m}(-1)^{|\alpha|-1}(|\alpha|-1)!z^\alpha/\alpha! \quad \text{for } m\geq 2 \end{bmatrix} .$$

Thus the values of μ on the indecomposable types are

$$\mu(\varepsilon((1^02^0..),P_1)) = -1 \quad , \quad \mu(\varepsilon((1^12^03^0..),P_1)) = 1$$
(4.15) and, for $m\geq 2$,
$$\mu(\varepsilon(\alpha,P_m)) = (-1)^{|\alpha|-1}(|\alpha|-1)! \quad \text{if } \omega(\alpha)=m \quad \text{and} \quad = 0 \quad \text{else} .$$

By Theorem (1.37) the Möbius function of $Sub(P_m)$, the lattice of partial partitions of $\{1,..,m\}$, can be computed. $Sub(P_m)$ satisfies the chain condition and possesses the rank function

$$r(\pi) = 2\#X - \#\pi \quad , \quad \pi \text{ a partition of a subset } X \text{ of } \{1,..,m\} \ .$$

In particular, the greatest element of $Sub(P_m)$, $\hat{1}=\{\{1,..,m\}\}$, has

rank 2m-1. We now set up a generating function for the Whitney numbers $W_m(i)$ of $Sub(P_m)$ ($W_m(i)$ = number of elements of $Sub(P_m)$ of rank i). Since the injection $(\emptyset,\emptyset)\to P_m$ has type $\varepsilon(0,P_m)$, we have $\Sigma_{i=0}^{2m-1} W_m(i)D^i = D^{\dim}\zeta(\varepsilon(0,P_m))$ by (1.37).
But the dimension function $\dim: T\to \mathbb{N}_0$ is given by $\dim(\varepsilon(\alpha,P_m)) = (2m-1)-(2\omega(\alpha)-|\alpha|) = 2(m-\omega(\alpha))+(|\alpha|-1)$ for $(\alpha,P_m)\in S$, and

$$\hat{\beta}(D^{\dim}) = \begin{bmatrix} w \to w \\ z_m \to \frac{1}{m!}(\partial^m/\partial x^m)|_{x=0}\ \{\exp(D^2wx+D\Sigma_{j=1}^{\infty}z_j)-1\}/D \end{bmatrix} .$$

Hence, applying $\hat{\beta}(D^{\dim})$ to $\Sigma_{m=1}^{\infty}\hat{\beta}(\zeta)(z_m) = \exp(w+\Sigma_{j=1}^{\infty}z_j)-1$, we obtain
$\Sigma_{m=1}^{\infty}\hat{\beta}(D^{\dim}\zeta)(z_m) = \exp(w+\{\exp(D^2w+D\Sigma_{j=1}^{\infty}z_j)-1\}/D)-1$,
and, setting $z_1=z_2=..=0$,

(4.16) $$\Sigma_{m=1}^{\infty}(\Sigma_{i=0}^{2m-1} W_m(i)D^i)\ w^m/m! = \exp(w+\{\exp(D^2w)-1\}/D)\ -1 .$$

In particular,

$$\Sigma_{m=1}^{\infty}\ \#Sub(P_m)\ w^m/m! = \exp(w+\exp(w)-1)\ -1 .$$

<u>Case 2:</u> $N=\mathbb{N}d=\{d,2d,3d,..\}$, $d\in\mathbb{N}$.

From (4.14) it follows by multisection of series ([Ri,68]) that

$$1+\Sigma_{m\in N}z_mx^m = \frac{1}{d}\Sigma_{r=0}^{d-1}\exp(\xi^r wx+\Sigma_{j\in N}\hat{\beta}(\mu)(z_j)x^j) =$$
$$= \exp(\Sigma_{j\in N}\hat{\beta}(\mu)x^j)\,(\tfrac{1}{d}\Sigma_{r=0}^{d-1}\exp(\xi^r wx))$$

where ξ is a primitive d-th root of unity in $\mathbb{C}$. Now, let

$$h_k(x,d) = \Sigma_{l=0}^{\infty}\ x^{dl+k-1}/(dl+k-1)! = \tfrac{1}{d}\Sigma_{r=0}^{d-1}\ \xi^{(1-k)r}\exp(\xi^r x) ,\quad k=1,..,d ,$$

be the <u>hyperbolic functions of order d</u> ([EMOT,55]), e.g.
$h_1(x,1)=\exp(x)$, $h_1(x,2)=\cosh(x)$ and $h_2(x,2)=\sinh(x)$.
By the above calculation,
$\Sigma_{j\in N}\hat{\beta}(\mu)(z_j)x^j = \log(1+\Sigma_{m\in N}z_mx^m)\ -\log(h_1(wx,d))$. Comparing coefficients, we get the values of μ on the indecomposable types:

(4.17)
$$\mu(\varepsilon(\alpha,P_m)) = -b_d(m/d) \quad \text{if } \alpha=(d^0 2d^0..) ,$$
$$= 0 \quad \text{if } 0<\omega(\alpha)<m , \text{ and}$$
$$= (-1)^{|\alpha|-1}(|\alpha|-1)! \quad \text{if } \omega(\alpha)=m .$$

Here the numbers $b_d(l)$ are defined implicitly by $\log(h_1(x,d)) = \Sigma_{l=1}^{\infty}\ b_d(l)x^{ld}/(ld)!$. For instance, if d=2, $b_d(l) = (-1)^l A_{2l-1}$ where the A_{2l-1} are the tangent numbers ([Co,74]). By Theorem (1.37), the Möbius function of the poset $Sub(P_m)$ takes the value

$$\mu_{Sub(P_m)}(\hat{0},\hat{1}) = -b_d(m/d)$$

where $\hat{0}=\emptyset$ is the least and $\hat{1}=\{\{1,..,m\}\}$ is the greatest element of $Sub(P_m)$. Since $\mu_{Sub(P_m)}(\pi,\hat{1})=0$ for any π where neither $\pi=\hat{0}$ nor $\pi\in\Pi_m^N$,

this implies a result of R.P.Stanley in [St,78']. In his article, Stanley considers the poset Q obtained from the poset Π_m^N by adjoining a least element $\hat{0}$. Generalizing results of G.S.Sylvester ([Sy,76]) and J.Rosen ([Ros,76]) in the special case d=2, he proves that
$\mu_Q(\hat{0},\hat{1}) = -b_d(m/d)$.

By a calculation similar to that in case 1, a generating function for the numbers $\#\mathrm{Sub}(P_m) = \zeta^2(\varepsilon(0,P_m))$ can be derived:

$$(4.18) \qquad \Sigma_{m\in N}\ \#\mathrm{Sub}(P_m)\ w^m/m! = \exp(h_1(w,d)-1)h_1(w,d)\ -1 \quad ,$$

in particular, for d=2,

$$\Sigma_{m\in N}\ \#\mathrm{Sub}(P_m)\ w^m/m! = \exp(\cosh(w)-1)\cosh(w)\ -1 \quad .$$

Case 3: $N = 1+\mathbb{N}_0 d = \{1,1+d,1+2d,..\}$, $d\geq 2$.

From (4.14) it follows that $\Sigma_{m\in N} z_m x^m =$
$= \frac{1}{d}\Sigma_{r=0}^{d-1}\ \xi^{-r}\exp(\xi^r wx+\Sigma_{j\in N}\hat{\beta}(\mu)(z_j)\xi^r x^j) = h_2(wx+\Sigma_{j\in N}\hat{\beta}(\mu)(z_j)x^j,d)$,
and hence $\Sigma_{j\in N}\hat{\beta}(\mu)(z_j)x^j = h_2^{-1}(\Sigma_{m\in N} z_m x^m,d)\ -wx$.

Defining numbers $c_d(l)$, $l\in\mathbb{N}_0$, by the expansion

$$h_2^{-1}(x,d) = \Sigma_{l=0}^{\infty}\ c_d(l)x^{ld+1}/(ld+1)! \quad ,$$

we can specify the values of μ on the indecomposable types:

$$\mu(\varepsilon((1^0 1{+}d^0..),P_1) = -1 \quad , \quad \mu(\varepsilon((1^1 1{+}d^0 1{+}2d^0..),P_1) = 1$$

(4.19) and, for $m\geq 2$,

$$\mu(\varepsilon(\alpha,P_m)) = c_d((|\alpha|-1)/d) \quad \text{if } \omega(\alpha)=m \quad \text{and} \quad = 0 \quad \text{else} .$$

In the special case d=2, the power series expansion of arsinh= $\sinh^{-1}$ shows that

$$c_2(l) = (-1)^l\ \Pi_{i=1}^{l}(2i-1)^2 \qquad (c_2(0)=1) \quad .$$

E.g., let d=2 and m=5. Here Π_m^N , the set of all partitions of {1,..,5} such that each block has odd size, contains
$\hat{1}$={{1,2,3,4,5}},
{{1},{2},{3,4,5}}, {{1},{3},{2,4,5}},..., {{4},{5},{1,2,3}}, and
$\hat{0}$={{1},{2},{3},{4},{5}},
while $\mu_{\Pi_m^N}(\hat{0},\hat{1}) = \mu(\varepsilon((1^5 3^0 5^0..),P_5) = c_2(2) = 9$ by (1.37).

Finally, by a calculation similar to that in case 1, we obtain

$$(4.20) \qquad \Sigma_{m\in N}\ \#\mathrm{Sub}(P_m)\ w^m/m! = h_2(w+h_2(w,d),d) \quad ,$$

in particular, for d=2,

$$\Sigma_{m\in N}\ \#\mathrm{Sub}(P_m)\ w^m/m! = \sinh(w+\sinh(w)) \quad .$$

§6. Partial partitions of finite vector spaces

Let $\mathbb{F}$ be a finite field with q elements. We consider the following category $\underline{K}$: The objects in $\underline{K}$ are pairs (X,π) where X is a vector space over $\mathbb{F}$ of finite dimension and where π is a partition of X in the category $\underline{Vf}_{\mathbb{F}}$ (Recall from (2.4),Example c, that a partition of X in $\underline{Vf}_{\mathbb{F}}$ is a finite set of non-zero subspaces of X such that X is the direct sum of them). A morphism from (X,π) to (Y,σ) is a linear function $s:X\to Y$ mapping each block of π into a block of σ.

In $\underline{K}$ the initial objects are the 0-dimensional vector spaces with the empty partition, and the direct sum of (X,π) and (Y,σ) is $(X\oplus Y,\pi\dot{\cup}\sigma)$.

An object (Z,τ) is indecomposable if and only if τ has only one block. Thus a system of representatives of the indecomposable objects in $\underline{K}$ modulo isomorphism is given by

$$\underset{\sim}{P} = \{P_m; m=1,2,..\} \qquad \text{where} \qquad P_m=(\mathbb{F}^m,\{\mathbb{F}^m\}) \quad .$$

The Krull-Schmidt type $n\in \mathbb{N}_o(\underset{\sim}{P})$ of an object (X,π) tells, for every m, how many blocks of π have dimension m. Instead of n, we use the "type" $\alpha\in \mathbb{N}_o(\mathbb{N})$ of the partition π which is written as $\alpha=(1^{\alpha(1)}2^{\alpha(2)}..)$. Then $\omega(\alpha)=\Sigma_{j=1}^{\infty} j\alpha(j)$ is the dimension of X whereas $|\alpha|=\Sigma_{j=1}^{\infty}\alpha(j)$ is the number of blocks of π. By (2.17) the automorphism group of (X,π) is isomorphic to the direct product of wreath products $\Pi_{j=1}^{\infty} Gl_{\mathbb{F}}(j) \text{ wr } S_{\alpha(j)}$ and has the order

$$a(\alpha) = \Pi_{j=1}^{\infty} b_q(j)^{\alpha(j)}\alpha(j)! \quad .$$

Here $Gl_{\mathbb{F}}(j)$ denotes the general linear group on $\mathbb{F}^j$ of order

$$b_q(j) = (q^j-1)(q^j-q)..(q^j-q^{j-1}) \quad .$$

Let M be the class of monomorphisms in $\underline{K}$. In $\underline{K}$ a morphism $s:(X,\pi)\to(Y,\sigma)$ is a monomorphism if and only if $s:X\to Y$ is one-to-one. By a <u>partial partition</u> of the vector space Y we understand a partition of a subspace of Y in $\underline{Vf}_{\mathbb{F}}$, i.e. a set of non-zero subspaces of Y whose sum is direct. Given two partial partitions π,τ of Y, π is called finer than τ (denoted by $\pi\leq\tau$) if each block of π is contained in a block of τ.

We identify a subobject $[r:(X,\pi)\to(Y,\sigma)]$ of (Y,σ) with the partial partition $\{r(B);B\in\pi\}$ of Y. Then $Sub(Y,\sigma)$ is the set of all partial partitions of Y finer than σ, ordered by refinement. The least element of $Sub(Y,\sigma)$ is $\emptyset$, the greatest element is σ.

Since $Sub(Y,\sigma)$ is order-isomorphic to the product $\Pi_{B\in\sigma} Sub(B,\{B\})$, it suffices to consider the posets $Sub(P_m)$ consisting of all partial partitions of $\mathbb{F}^m$, $m\in \mathbb{N}$. Each $Sub(P_m)$ is a finite point lattice which satisfies the Jordan-Dedekind chain condition with the rank function

$$r(\pi) = 2\dim_{\mathbb{F}}X - \#\pi \quad \text{where} \quad X=\oplus_{\pi}B \quad .$$

In particular, the rank of the greatest element $\hat{1}=\{\mathbb{F}^m\}$ is 2m-1.

Finally, let $\sim$ be the isomorphism relation on M. The isomorphism type of an indecomposable monomorphism $s:(X,\pi)\to(Y,\{Y\})$ is uniquely determined by the KS-type α of (X,π) and the dimension m of Y. Since s is one-to-one, $\dim_{\mathbb{F}}X=\omega(\alpha)\leq m=\dim_{\mathbb{F}}Y$. Thus we are in the situation of (3.44) with

$$S=\{(\alpha,P_m);\alpha\in\mathbb{N}_o(\mathbb{N}),m\in\mathbb{N}\text{ such that }\omega(\alpha)\leq m\}\quad\text{and}\quad T=\mathbb{N}_o(S)\quad .$$

To compute the elementary section coefficients (3.45), we choose a vector space Y over $\mathbb{F}$ of dimension m and a partition π of a subspace X in $\underline{\mathrm{Vf}}_{\mathbb{F}}$ such that (X,π) has KS-type α. Then the injection $s:(X,\pi)\to(Y,\{Y\})$ is a representative of $\varepsilon(\alpha,P_m)$, and by (1.33), $G(\varepsilon(\alpha,P_m);t,\varepsilon(c(t),P_m))=\#A(Y,\pi,t)$ where $A(Y,\pi,t)$ is the set of all partitions τ of some subspace Z of Y in $\underline{\mathrm{Vf}}_{\mathbb{F}}$ which are coarser than π and such that the injection $(X,\pi)\to(Z,\tau)$ has type t. Let G be the group of all linear transformations of Y mapping X to itself as an automorphism of (X,π), and consider the operation of G on $A(Y,\pi,t)$ given by $g\tau=\{g(B);B\in\tau\}$ for $g\in G$ and $\tau\in A(Y,\pi,t)$. Evidently, the order of G is $a(\alpha)b_q(m-\omega(\alpha))q^{\kappa(\alpha,m)}$ where $\kappa(\alpha,m)=\omega(\alpha)(m-\omega(\alpha))$. Furthermore, it is not difficult to see that G acts transitively on $A(Y,\pi,t)$ and that the stabilizer of any $\tau\in A(Y,\pi,t)$ has the order $o(m;t)=$ $=(\Pi_{(\beta,n)\in S}(a(\beta)b_q(n-\omega(\beta))q^{\kappa(\beta,n)})^{t(\beta,n)}t(\beta,n)!)b_q(m-\omega(c(t)))q^{\kappa(c(t),m)}$.

We conclude that

$$G(\varepsilon(\alpha,P_m);t,\varepsilon(c(t),P_m))=\frac{a(\alpha)b_q(m-\omega(\alpha))q^{\kappa(\alpha,m)}}{o(m;t)}\quad .$$

Now, for every $l\in\mathbb{N}_o$, define the $\underline{\text{q-factorial}}$ of l by

$$[l]!=\Pi_{i=1}^{l}[i]\quad,\quad[i]=q^{i-1}+q^{i-2}+..+q+1\qquad(\ [0]!=1\ )\quad .$$

Then $b_q(l)=[l]!(q-1)^lq^{l(l-1)/2}$. Using the multiplicative function $\varphi:T\to E(\mathbb{Q})$, $\varphi(\varepsilon(\alpha,P_m))=\alpha![m-\omega(\alpha)]!q^{\lambda(m,\omega(\alpha))}$ for $(\alpha,P_m)\in S$ where

$$\lambda(m,l)=(m-l)(m+l-1)/2\quad,$$

this can be written as $c(t)!\varphi(\varepsilon(\alpha,P_m))/t!\varphi(t)\varphi(\varepsilon(c(t),P_m))$.
By Proposition (3.46) we obtain the following result.

(4.21) THEOREM: The affine monoid of multiplicative functions on T has the faithful representation

$$\begin{array}{llll}\rho_R: & Mu(R) & \to & End(R[z_m;m=1,2,..])\qquad,\ R\in Al_{\mathbb{Q}}\ ,\\ & f & \to & [z_m\to\Sigma_\alpha f(\varepsilon(\alpha,P_m))z^\alpha/\alpha![m-\omega(\alpha)]!q^{\lambda(m,\omega(\alpha))}]\end{array}$$

by endomorphisms of the polynomial algebra $\mathbb{Q}[z_m;m=1,2,..]$. □

As in §5, we shall use a power series representation of Mu induced by ρ.

(4.22) COROLLARY: The affine monoid of multiplicative functions on T admits the faithful power series representation

$$\hat{\rho}_R: \mathrm{Mu}(R) \to \mathrm{End}(R[[w,z_m;m=1,2,..]]) \quad , \quad R\in\mathrm{Al}_{\mathbb{Q}} ,$$

$$f \to \begin{bmatrix} w \to w \\ z_m \to \Sigma_{l=0}^{m} q^{-\lambda(m,l)} (\Sigma_{\alpha,\omega(\alpha)=l} f(\varepsilon(\alpha,P_m))\frac{z^\alpha}{\alpha!}) \frac{w^{m-l}}{[m-l]!} \end{bmatrix} .$$

□

Under this representation, the zeta function $\zeta\in\mathrm{Mu}"(\mathbb{Z})$ acts by the transformation

$$\hat{\rho}(\zeta) = \begin{bmatrix} w \to w \\ z_m \to \Sigma_{l=0}^{m} q^{-\lambda(m,l)} \{\frac{1}{l!}(\partial^l/\partial x^l)|_{x=0}\exp(\Sigma_{j=1}^{\infty} z_j x^j)\} \frac{w^{m-l}}{[m-l]!} \end{bmatrix}$$

where x is a new variable and $(\partial^l/\partial x^l)|_{x=0}$ means differentiation followed by setting x=0. Our next aim is to compute the Möbius function $\mu=\zeta^{-1}\in\mathrm{Mu}"(\mathbb{Z})$. Since $\hat{\rho}(\mu)= \hat{\rho}(\zeta)^{-1}$, we have

$$\Sigma_{l=0}^{m} q^{-\lambda(m,l)} s_l w^{m-l}/[m-l]! = z_m \quad , \quad m\in \mathbb{N} ,$$

where $s_l = \frac{1}{l!}(\partial^l/\partial x^l)|_{x=0}\exp(\Sigma_{j=1}^{\infty}\hat{\rho}(\mu)(z_j)x^j)$. Putting $z_0=1$, these equations can be written as

$$\Sigma_{l=0}^{m} \begin{bmatrix} m \\ l \end{bmatrix} ([l]!q^{l(l-1)/2} s_l/w^l) = [m]!q^{m(m-1)/2} z_m/w^m \quad , \quad m\in \mathbb{N}_o ,$$

where the numbers $\begin{bmatrix} m \\ l \end{bmatrix} = \frac{[m]!}{[l]![m-l]!}$ are the Gaussian coefficients ([Ai,79]). Applying Gauss inversion, we obtain

$$\Sigma_{l=0}^{m} z_l(-w/q^l)^{m-l}/[m-l]! = s_m \quad , \quad m\in \mathbb{N}_o ,$$

and, after a little calculation,

$$\exp(\Sigma_{j=1}^{\infty}\hat{\rho}(\mu)(z_j)x^j) = \Sigma_{m=0}^{\infty} s_m x^m = (1+\Sigma_{m=1}^{\infty} z_m x^m/p_m(wx))e(-wx) .$$

Here $p_m(x)=(1-(q-1)x/q^m)(1-(q-1)x/q^{m-1})..(1-(q-1)x/q)$, and $e(x)= \Sigma_{i=0}^{\infty} x^i/[i]!$ is the q-exponential function ([Ci,82]). Observing that $\log(e(-x))= -\Sigma_{m=1}^{\infty} (q-1)^m/m(q^m-1)\, x^m$, we have established

(4.23) PROPOSITION: The power series $\hat{\rho}(\mu)(z_m)$, $m\in \mathbb{N}$, possess the generating function

$$\Sigma_{m=1}^{\infty} \hat{\rho}(\mu)(z_m)x^m = \log(1+\Sigma_{m=1}^{\infty} z_m x^m/p_m(wx)) - \Sigma_{m=1}^{\infty} (q-1)^m/m(q^m-1)\,(wx)^m$$

where the $p_m(x)$ are the polynomials

$$(1-(q-1)x/q^m)(1-(q-1)x/q^{m-1})..(1-(q-1)x/q) \quad . \quad □$$

Expanding log(1+..) into a power series and comparing coefficients, the values of μ on the indecomposable types can be determined.

(4.24) COROLLARY: Let $(\alpha, P_m) \in S$.
If $\alpha=0$,

$$\mu(\varepsilon(\alpha,P_m)) = -\frac{1}{m}(q-1)^{m-1} q^{m(m-1)/2}[m-1]! \quad .$$

If $\alpha \neq 0$,

$$\mu(\varepsilon(\alpha,P_m)) = (-1)^{|\alpha|-1}(|\alpha|-1)!(q-1)^{d}[d]!\ Q_{\alpha,m}(q)$$

where $d=m-\omega(\alpha)$ and the polynomial $Q_{\alpha,m}(q)$ is given as follows: Let l be the largest j such that $\alpha(j) \neq 0$, and define $\beta \in \mathbb{N}_o(\mathbb{N})$ by $\beta(m) = \Sigma_{j=m}^{\infty} \alpha(j)$ for $m \in \mathbb{N}$. Then

$$Q_{\alpha,m}(q) = \Sigma \left\{\Pi_{j=1}^{l} \binom{\beta(j)+\gamma(j)-1}{\gamma(j)}\right\} q^{\lambda(m,\omega(\alpha))-\omega(\gamma)} \quad ,$$

the sum running over all $\gamma \in (\mathbb{N}_o)^l$ such that $\Sigma_{j=1}^{l}\gamma(j)=d$. Here $\binom{\beta(j)+..}{\gamma(j)}$ is the binomial coefficient, and $\omega(\gamma) = \Sigma_{j=1}^{l} j\gamma(j)$. □

By Theorem (1.37) we can compute the Möbius function of the lattice of partial partitions of $\mathbb{F}^m$. For instance,

$$\mu_{Sub(P_m)}(\hat{0},\hat{1}) = -\frac{1}{m}\ \Pi_{i=1}^{m-1}(q^m-q^i)$$

where $\hat{0}=\emptyset$ is the least and $\hat{1}=\{\mathbb{F}^m\}$ is the greatest element of $Sub(P_m)$. In some sense, the lattice of partial partitions of $\mathbb{F}^m$ is a <u>q-analogue</u> of the lattice of partial partitions of finite sets studied in §5, Case 1: All the results proved here specialize to the results obtained there when formally putting $q=1$, in particular the power series representation $\hat{\beta}$. For the lattices of subsets of finite sets and the lattices of subspaces of finite vector spaces, this phenomenon was pointed out by J.Goldman and G.C.Rota in [GR,70].

As an application of Möbius inversion on the lattices $Sub(P_m)$ we now determine the distribution of the number of eigenvalues in $\mathbb{F}$ of an $m\times m$-matrix whose entries are chosen from $\mathbb{F}$ at random. Remarkably, it turns out that the expected number of eigenvalues tends to 1 when the field becomes large, for every m.

(4.25) THEOREM: For $m \in \mathbb{N}$ and $0 \leq k \leq m$, let $N(m,q;k)$ denote the number of $m\times m$-matrices with entries in $\mathbb{F}$ which have exactly k eigenvalues in $\mathbb{F}$. Then, in $\mathbb{Q}[z][[x]]$,

$$1 + \Sigma_{m=1}^{\infty} \Sigma_{k=0}^{m} N(m,q;k) z^k\ x^m/q^{m(m-1)/2}[m]! = \frac{\{e(-x)+z(1-e(-x))\}^q}{(1-(q-1)x)e(-x)}$$

where $e(x) = \Sigma_{i=0}^{\infty} x^i/[i]!$ is the q-exponential function ([Ci,82]).

We postpone the proof in order to draw some conclusions.

(4.26) PROPOSITION: Let $m \in \mathbb{N}$. Then the average number of eigenvalues in $\mathbb{F}$ of an $m\times m$-matrix over $\mathbb{F}$ is

$$\bar{N}(m,q) = \Sigma_{n=1}^{m} (1-q^{-m})(1-q^{-(m-1)})\dots(1-q^{-(n+1)})/q^{n-1}$$

(the last summand being $1/q^{m-1}$).

In particular, $\bar{N}(m,q) \to 1$ when $q \to \infty$ (independently of m).

For instance, $\bar{N}(1,q) = 1$ and $\bar{N}(2,q) = (q^2+q-1)/q^2$.

Proof: To compute $\bar{N}(m,q) = \Sigma_{k=0}^{m} kN(m,q;k)/q^{m^2}$, we differentiate in (4.25) with respect to z and subsequently set z=1, which gives

$\Sigma_{m=1}^{\infty} \bar{N}(m,q)q^{(m+1)m/2}x^m/[m]! = q(1-(q-1)x)^{-1}(\frac{1}{e(-x)} -1)$.

As $\frac{1}{e(-x)} = \Sigma_{n=0}^{\infty} q^{n(n-1)/2}x^n/[n]!$ (cf. (1.54)), the result follows. □

(4.27) PROPOSITION: For every $m \in \mathbb{N}$, let $z_q(m) = N(m,q;0)$ be the number of $m\times m$-matrices over $\mathbb{F}$ having no eigenvalue in $\mathbb{F}$. Then the numbers $z_q(m)$, $m \in \mathbb{N}$, possess the generating function

$$1 + \Sigma_{m=1}^{\infty} z_q(m)\, x^m/q^{m(m-1)/2}[m]! = \frac{e(-x)^{q-1}}{1-(q-1)x} ,$$

and can be computed recursively by the formula

$$z_q(n+1) = q^{n+1}(q^n-1)z_q(n) -$$
$$- q^n \Sigma_{i=1}^{n} (-1)^i \binom{q}{i+1} q^{in-(i+1)i/2}(q^n-1)\dots(q^{n-i+1}-1)z_q(n-i) ,$$

starting with $z_q(0) = 1$. For example,

$z_q(1) = 0$,

$z_q(2) = (q-1)^2q^2/2$,

$z_q(3) = (q-1)^3q^4(q+1)^2/3$,

$z_q(4) = (q-1)^4q^7(q^2+q+1)(3q^3+4q^2+5q+2)/8$.

In general, $z_q(m)$ has the form $R_m(q)/m!$ where R_m is a polynomial in one variable X with integer coefficients. If $m\geq 2$, then R_m has the degree m^2, the divisor $(X-1)^m X^m$ and the leading coefficient

$$m!\, \Sigma_{i=0}^{m} (-1)^i/i! = d(m)$$

which is the m-th derangement number ([Co,74]). In particular,

$$\lim_{q\to\infty} z_q(m)/q^{m^2} = d(m)/m! = \lim_{q\to\infty} z_q(m)/b_q(m) .$$

Here $z_q(m)/q^{m^2}$, $d(m)/m!$ and $z_q(m)/b_q(m)$ can be interpreted as the probabilities that an $m\times m$-matrix over $\mathbb{F}$ has no eigenvalue in $\mathbb{F}$, that a permutation of m elements leaves no element fixed and that a projective transformation in the (m-1)-dimensional projective space over $\mathbb{F}$ has no fixed point, respectively.

On the other hand,

$$\lim_{m\to\infty} z_q(m)/q^{m^2} = e(-\frac{1}{q-1})^q = \Pi_{n=1}^{\infty}(1-q^{-n})^q ,$$

e.g. for the binary field $\mathbb{F}=\{0,1\}$ the limit is approximately 0.083 .

Proof: We set z=0 in (4.25) and obtain

$$(*)\qquad 1+\Sigma_{m=1}^{\infty}\, z_q(m)\; x^m/q^{m(m-1)/2}[m]! = \frac{e(-x)^{q-1}}{1-(q-1)x}\ .$$

Then, applying the q-derivative D, $(Dh)(x)=\frac{h(qx)-h(x)}{(q-1)x}$ for $h\in\mathbb{C}[[x]]$, to both sides of (*), the right side appears again and can be replaced by the left side of (*). Comparing coefficients, we get the recursion formula for the $z_q(m)$, $m\in\mathbb{N}$. From this, the assertions on the shape of $z_q(m)$ follow easily by induction. Substituting x by x/(q-1) in (*) gives $1+\Sigma_{m=1}^{\infty} z_q(m)/b_q(m)\; x^m = e(-\frac{x}{q-1})^{q-1}/(1-x)$. Because of the simple pole at 1, $\lim_{m\to\infty} z_q(m)/b_q(m) = e(-\frac{1}{q-1})^{q-1}$. Further, $b_q(m)/q^{m^2} = \Pi_{i=1}^{m}(1-q^{-i})$ and $\Pi_{i=1}^{\infty}(1-q^{-i}) = e(-\frac{1}{q-1})$ by an Eulerian identity ([An,76]). Combining these results, we conclude that $\lim_{m\to\infty} z_q(m)/q^{m^2} = e(-\frac{1}{q-1})^q$. □

(4.28) REMARK: A similar, but more complicated expression for the generating function of the numbers $z_q(m)$, $m\in\mathbb{N}$, has been obtained by J.P.S.Kung in [Ku,81] where a vector space analogue of the Pólya cycle index is introduced.

If q=2, then $z_2(m)/b_2(m) = \Sigma_{i=0}^{m}(-1)^i/[i]!$. The numbers

$$D_n(q) = [n]!\; \Sigma_{j=0}^{n}(-1)^j/[j]!$$

have been studied by A.M.Garsia and J.Remmel ([GaRe,80]) as a q-analogue of the derangement numbers d(n). For arbitrary q however, $z_q(m)/b_q(m) \neq D_m(q)/[m]!$ in general, e.g. when q=3 and m=2.

Finally, it can be shown that, as $q\to 1$, $z_q(m)/b_q(m)$ tends to the coefficient of x^m in the powerseries $\exp(-\Sigma_{n=1}^{\infty} x^n/n^2)/(1-x)$. Since $\Sigma_{m=1}^{\infty} d(m)/m!\; x^m = \exp(-x)/(1-x)$, we see that in general $z_q(m)/b_q(m)$ does not converge to d(m)/m! when $q\to 1$.

Proof of Theorem (4.25): Let $m\in\mathbb{N}$. We denote the set of m×m-matrices with entries in $\mathbb{F}$ by $M_m(\mathbb{F})$ and view a matrix in $M_m(\mathbb{F})$ as a linear operator on $\mathbb{F}^m$. For $\pi\in\mathrm{Sub}(P_m)$, let $A_m(\pi)$ be the set of all $C\in M_m(\mathbb{F})$ such that π is the partial partition of $\mathbb{F}^m$ into eigenspaces of C, and let $B_m(\pi)$ be the set of all $C\in M_m(\mathbb{F})$ such that C acts on each block of π as a scalar multiple of the identity. Then, for all $\pi\in\mathrm{Sub}(P_m)$, $B_m(\pi)$ is the disjoint union of those $A_m(\sigma)$ where $\sigma\geq\pi$ in $\mathrm{Sub}(P_m)$. Therefore,

$$\#B_m(\pi) = \Sigma_{\sigma\geq\pi}\; \#A_m(\sigma) \qquad \text{for all } \pi\in\mathrm{Sub}(P_m)\ .$$

But the numbers $\#A_m(\pi)$ and $\#B_m(\pi)$ depend only on the type $\varepsilon(\alpha,P_m)\in U$ where α denotes the KS-type of (X,π), $X=\oplus_{\pi}B$. Thus we can define two multiplicative functions $f,g:T\to\mathbb{Z}$ by $f(\varepsilon(\alpha,P_m)) = \#A_m(\pi)$ and

$g(\varepsilon(\alpha,P_m)) = \#B_m(\pi)$. Evidently, $g(\varepsilon(\alpha,P_m)) = q^{|\alpha|+m(m-\omega(\alpha))}$ because π contains exactly $\alpha(j)$ blocks of dimension j, for every j.
The preceding considerations imply that

$g(\varepsilon(\alpha,P_m)) = \Sigma_{t\in T}\, G(\varepsilon(\alpha,P_m);t,\varepsilon(c(t),P_m))f(\varepsilon(c(t),P_m)) = \zeta f(\varepsilon(\alpha,P_m))$

since, by (1.33), the section coefficients count subobjects.
We hence have $g=\zeta f$ and $f=\mu g$ in $Mu(\mathbb{Z})$, the last step corresponding to a Möbius inversion on the lattices $Sub(P_m)$.

From $\hat{\beta}(g)(z_m) = \Sigma_{l=0}^{m}\, q^{-\lambda(m,l)}(\Sigma_{\alpha,\omega(\alpha)=l}\, q^{|\alpha|+m(m-l)}\frac{z^\alpha}{\alpha!})\frac{w^{m-l}}{[m-l]!}$, $m\in\mathbb{N}$,

it follows by a little calculation that

$$1+\Sigma_{m=1}^{\infty}\hat{\beta}(g)(z_m)x^m = \frac{1}{(1-(q-1)wx)e(-wx)}\exp(\Sigma_{j=1}^{\infty}z_jx^j)^q$$

where x is a new variable. We now apply $\hat{\beta}(\mu)$, recalling that

$\exp(\Sigma_{j=1}^{\infty}\hat{\beta}(\mu)(z_j)x^j) = (1+\Sigma_{m=1}^{\infty}z_mx^m/p_m(wx))e(-wx)$, to obtain

$$(**)\qquad 1+\Sigma_{m=1}^{\infty}\hat{\beta}(f)(z_m)x^m = (1+\Sigma_{m=1}^{\infty}z_mx^m/p_m(wx))^q\,\frac{e(-wx)^{q-1}}{1-(q-1)wx}\ .$$

This is the generating function for the numbers $f(\varepsilon(\alpha,P_m))$, $(\alpha,P_m)\in S$, which will be used to set up the generating function for the numbers $N(m,q;k)$, $0\leq k\leq m$. First,

$N(m,q;k) = \Sigma_{\alpha,|\alpha|=k}\begin{bmatrix} m\\ \omega(\alpha)\end{bmatrix} b_q(\omega(\alpha))/a(\alpha)\ f(\varepsilon(\alpha,P_m))$

because there are $\begin{bmatrix} m\\ \omega(\alpha)\end{bmatrix} b_q(\omega(\alpha))/a(\alpha)$ partial partitions π of $\mathbb{F}^m$ such that $(\oplus_\pi B,\pi)$ has the KS-type α, according to Proposition (2.8).

Consequently, $\Sigma_{k=0}^{m}N(m,q;k)z^k =$

$= q^{m(m-1)/2}[m]!\{\Sigma_\alpha\, q^{-\lambda(m,\omega(\alpha))}f(\varepsilon(\alpha,P_m))\frac{(q-1)^{\omega(\alpha)}z^{|\alpha|}}{a(\alpha)}\ \frac{1}{[m-\omega(\alpha)]!}\}$

where z is a new variable. But the expression on the right side can be obtained from $q^{m(m-1)/2}[m]!\ \hat{\beta}(f)(z_m)$ by putting w=1 and, for all j, $z_j = z/q^{j(j-1)/2}[j]!$. With these substitutions, (**) becomes

$$1+\Sigma_{m=1}^{\infty}\Sigma_{k=0}^{m}\ N(m,q;k)z^kx^m/q^{m(m-1)/2}[m]! = \frac{\{e(-x)+z(1-e(-x))\}^q}{(1-(q-1)x)e(-x)}$$

because, when setting $z_j = z/q^{j(j-1)/2}[j]!$ and w=1 in the equation

$(1+\Sigma_{m=1}z_mx^m/p_m(wx))e(-wx) = 1+\Sigma_{m=1}^{\infty}\{\Sigma_{l=0}^{m}z_l(-w/q^l)^{m-l}/[m-l]!\}x^m\qquad(z_0=1)$

established earlier, the right side reduces to $e(-x)+z(1-e(-x))$. □

(4.29) REMARK: Obviously, the same arguments apply when considering invertible matrices instead of arbitrary ones. If $I(m,q;k)$ denotes the number of invertible m×m-matrices with entries in $\mathbb{F}$ having exactly k eigenvalues in $\mathbb{F}-\{0\}$, it can be proved in a similar way that

$$1+\Sigma_{m=1}^{\infty}(\Sigma_{k=0}^{m}\ I(m,q;k)/b_q(m)\ z^k)\ x^m = \frac{1}{1-x}\{e(-\frac{x}{q-1})+z(1-e(-\frac{x}{q-1}))\}^{q-1}$$

in $\mathbb{Q}[z][[x]]$. Using this, one can study the distribution of the number of eigenvalues in $\mathbb{F}$ on the general linear groups over $\mathbb{F}$, e.g. determine its moments and derive asymptotic results.

§7. A class of geometric lattices based on finite groups

In this section we show how the geometric lattices of T.A.Dowling ([Dow,73'],[Han,84]) can be studied within our general theory. Whereas Dowling has derived the Möbius function, the characteristic polynomial and a recursion formula for the Whitney numbers by combinatorial arguments, we shall compute them from a power series representation by algebraic means. Our presentation is inspired by the approach of P.Doubilet, G.C.Rota and R.Stanley in a special case ([DRS,72]; compare also [Dow,73]).

Let G be a finite group of order m, and define a category $\underline{K}$ as follows: An object in $\underline{K}$ is a pair (X,α) where X is a finite set and where $\alpha=\{a_i:A_i\to G;i=1,..,l\}$ is a collection of functions into G for which the domains A_i are disjoint non-empty subsets of X. Following Dowling, we call α a partial G-partition of X with blocks A_i.

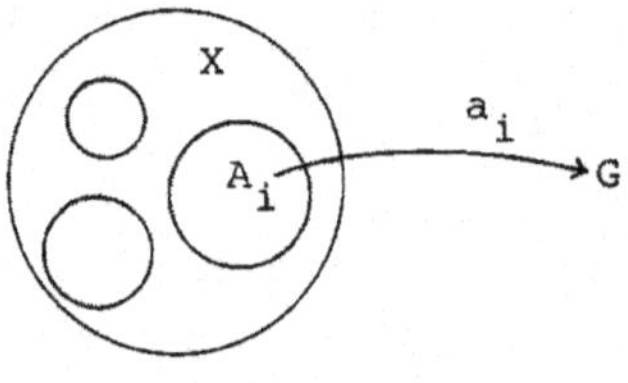

A morphism from one object (X,α) to another object (Y,β) is a map $s:X\to Y$ such that, for each β-block B_j, the inverse image $s^{-1}(B_j)$ is the union of some α-blocks A_i, and $b_j\circ s$ is a left-multiple of a_i on each block A_i (i.e. there exists a $\lambda_i\in G$ such that $b_j(s(x))=\lambda_i a_i(x)$ for all $x\in A_i$).
In $\underline{K}$ the initial object $0=(\emptyset,\emptyset)$ is the empty set with the empty partial G-partition, and the direct sum $(X,\alpha)\oplus(Y,\beta)=(X\dot{\cup}Y,\alpha\dot{\cup}\beta)$ is formed by disjoint union. A morphism $s:(X,\alpha)\to(Y,\beta)$ is a monomorphism if and only if $s:X\to Y$ is one-to-one. The indecomposable objects in $\underline{K}$ are the $(X,\emptyset)$ where $\#X=1$, as well as the $(X,\{a:X\to G\})$ where X is not empty.
An object (X,α) has the Krull-Schmidt decomposition
$(X,\alpha)=(\oplus_{x\in A_o}(\{x\},\emptyset))\oplus(\oplus_{i=1}^{l}(A_i,\{a_i:A_i\to G\}))$ where $A_o=X-\cup_{i=1}^{l}A_i$.
Thus $\underline{K}$ is a KS-category with unique KS-partitions.

We next introduce the geometric lattices of Dowling. Let Y be a finite set. On the set $Q_Y'(G)$ of all partial G-partitions of Y, a preorder $\leq$ is given by

$\alpha\leq\beta$ if and only if each β-block B_j is the union of some α-blocks A_i, and b_j is a left-multiple of a_i on each block A_i.

Obviously, $\alpha\leq\beta$ and $\beta\leq\alpha$ if and only if α,β contain the same functions, up to left-multiples. Let $Q_Y(G)$ be the set of equivalence classes of partial G-partitions of Y with respect to this equivalence relation, and let (α) denote the equivalence class of α. Then the preorder $\leq$ on $Q_Y'(G)$ induces a partial order $\leq$ on $Q_Y(G)$: $(\alpha)\leq(\beta)$ if and only if $\alpha\leq\beta$.
In the finite poset $Q_Y(G)$, the greatest element is $\hat{1}=(\emptyset)$ while

the least element is $\hat{0}=(\varepsilon)$, $\varepsilon=\{e_y;y\in Y\}$ where the $e_y:\{y\}\to G$, $e_y(y)=1$, are the unit functions.
In [Dow,73'], Dowling shows that $Q_Y(G)$ is a geometric lattice of rank $r=\#Y$ with considerable structure. For instance, $Q_Y(G)$ possesses a Boolean sublattice of modular elements and so is supersolvable ([St,72]). If G is the trivial group $\{1\}$, then $Q_Y(\{1\})$ is isomorphic to the partition lattice of the set $Y\cup\{y_o\}$, y_o being a new element, by $(\alpha=\{a_i:A_i\to\{1\};i=1,..,l\})\to\{A_o\cup\{y_o\},A_1,..,A_l\}$ where $A_o=Y-\cup_{i=1}^{l}A_i$.
Furthermore, the lattice $Q_Y(G)$ does reflect the structure of the underlying group G since Dowling proves:
If Y has at least 3 elements and $Q_Y(G)$ is isomorphic to $Q_Y(G')$ for another group G', then G' is isomorphic to G.

In the category $\underline{K}$ these lattices appear naturally. Let M be the class of all monomorphisms $s:(X,\alpha)\to(Y,\beta)$ where X and Y have the same cardinality. A subobject $[r:(X,\alpha)\to(Y,\beta)]$ of (Y,β) in M can be identified with the equivalence class $(s(\alpha))$ of the partial G-partition $s(\alpha)=$ $=\{a_i\circ s^{-1}:s(A_i)\to G;i=1,..,l\}$ where $\alpha=\{a_i:A_i\to G;i=1,..,l\}$. Then

$$\mathrm{Sub}_M(Y,\beta)=\{(\alpha)\in Q_Y(G);(\alpha)\leq(\beta)\}\quad,$$

and the order of $\mathrm{Sub}_M(Y,\beta)$ coincides with the order induced from $Q_Y(G)$.
In particular, $Q_Y(G)=\mathrm{Sub}_M(Y,\emptyset)$.
M clearly has the properties (M1)-(M5). Condition (M6) reads:
If $s:(X,\alpha)\to(Y,\beta)$ is a morphism in M and if $\pi=\{[r_k:(Y_k,\beta_k)\to(Y,\beta)];k=1,2\}$ is a partition of (Y,β), then there exists a unique partition σ of s such that π is the image partition of (Y,β) induced by σ.
However, this fails in general, e.g. when $\alpha=\{a:X\to G\}$ and $\beta=\emptyset$, but holds under the assumption that the union of all β_2-blocks is Y_2.
Now, for $s:(X,\alpha)\to(Y,\beta)\in M$, let $\tau(s)\in\mathbb{N}_o(\mathbb{N}_o)$ be given by

$\tau(s)(0)=$ number of α-blocks which are mapped into no β-block ,

and, for $n\geq1$, by

$\tau(s)(n)=$ number of β-blocks which contain the images of exactly n α-blocks.

Then it is easy to verify that the equivalence relation $\sim$ on M defined by $s_1\sim s_2$ if and only if $\tau(s_1)=\tau(s_2)$, satisfies the conditions (~1)-(~6). By the identification $\bar{s}=\tau(s)$ we have

$$T=\mathbb{N}_o(\mathbb{N}_o)\qquad\text{and}\qquad U=\{\varepsilon(n);n\in\mathbb{N}_o\}\quad.$$

We conclude that the triple $\underline{K}$,M,$\sim$ constitutes a categorical structure which is nearly sheaflike. Thus we can apply the theory of incidence algebras from chapter I. However, the incidence algebra $\mathbb{Z}(T;G)$ is not a bialgebra with the comultiplication induced from the addition in T. Nevertheless, it is possible to derive an affine monoid which contains the combinatorically important functions ζ,μ and $D^{\dim}$ and also admits a faithful power series representation.

We proceed as in chapter III. The relaxed condition (M6) yields an identity for the section coefficients:
For $v,w\in T$ with $w(0)=0$ and $t_1,t_2\in T$,

$$G(v+w;t_1,t_2) = \Sigma\, G(v;v_1,v_2)G(w;w_1,w_2)$$

where the summation is taken over all types $v_1,v_2,w_1,w_2\in T$ such that $v_1+w_1=t_1$ and $v_2+w_2=t_2$.
Consequently, $\dim(v+w) = \dim(v)+\dim(w)$ for $v,w\in T$ with $w(0)=0$.

(4.30) DEFINTTION: Let R be a commutative $\mathbb{Z}$-algebra with unit 1.
A function $f:T\to R$ is called <u>nearly multiplicative</u> if

$$f(0)=1 \quad \text{and} \quad f(v+w) = f(v)f(w) \quad \text{for all } v,w\in T \text{ with } w(0)=0 .$$

Then f is given by its values on the types $r\varepsilon(0)$, $r\in\mathbb{N}$, and $\varepsilon(n)$, $n\in\mathbb{N}$:

$$f(t) = f(t(0)\varepsilon(0))\ \Pi_{n=1}^{\infty} f(\varepsilon(n))^{t(n)} \quad \text{for } t\in T \ .$$

As expected, the set Mun(R) of nearly multiplicative functions on T with values in R is a submonoid of the incidence algebra R(T;G) with respect to the convolution multiplication: For $v,w\in T$ with $w(0)=0$,

$$\begin{aligned} fg(v+w) &= \Sigma_{t_1,t_2\in T}\, G(v+w;t_1,t_2)f(t_1)g(t_2) = \\ &= \Sigma_{v_1,v_2,w_1,w_2\in T}\, G(v;v_1,v_2)G(w;w_1,w_2)f(v_1+w_1)g(v_2+w_2) = \\ &= fg(v)fg(w) \end{aligned}$$

because $G(w;w_1,w_2)=0$ unless $w_1(0)=0=w_2(0)$. Hence,

(4.31) $\qquad \text{Mun}: \underline{\text{Al}}_{\mathbb{Z}} \to \underline{\text{Mo}}\ ,\ R \to \text{Mun}(R)\ ,$

defines a free affine monoid over $\mathbb{Z}$ which is represented by the polynomial algebra $\mathbb{Z}[x(r\varepsilon(0)),x(\varepsilon(n));r,n\in\mathbb{N}]$. We call Mun the <u>affine monoid of nearly multiplicative functions</u>. Its group of invertible elements E(Mun) inherits the algebraic structure of the group Ω described in Theorem (1.23), and thus is triangulable. Obviously, $\zeta,\mu\in\text{Mun}(\mathbb{Z})$ while $D^{\dim}\in\text{Mun}(\mathbb{Z}[D])$.

To compute the convolution product of $f,g\in\text{Mun}(R)$, one only has to know the section coefficients

$$G(r\varepsilon(0);t_1,t_2) \quad\text{and}\quad G(\varepsilon(n);t_1,t_2) \quad ,\quad r,n\in\mathbb{N},\ t_1,t_2\in T \quad ,$$

which count equivalence classes of partial G-partitions by (1.33).
Looking at the lattices $Q_Y(G)$, we find that

$$G(r\varepsilon(0);t,(|t|-t(0))\varepsilon(0)) = r!m^{r-|t|}/\Pi_{k\geq 0}(k!)^{t(k)}t(k)!$$

where $t\in T$ such that $t(0)+\Sigma_{k\geq 1}kt(k)=r$, and

$$G(\varepsilon(n);t,\varepsilon(|t|)) = n!/\Pi_{k\geq 1}(k!)^{t(k)}t(k)!$$

where $t\in T$ such that $t(0)=0$ and $\Sigma_{k\geq 1}kt(k)=n$.
Here m is the order of G, and $|t|=\Sigma_{k\geq 0}t(k)$.

(4.32) THEOREM: The affine monoid of nearly multiplicative functions on T has the faithful power series representation

$$\rho_R: \mathrm{Mun}(R) \to \mathrm{End}(R[[w,z]]) \quad , \ R \in \mathrm{Al}_{\mathbb{Q}} \ ,$$
$$f \to \begin{bmatrix} w \to \Sigma_{n=1}^{\infty} f(\varepsilon(n)) w^n/n! \\ z \to z(1+ \Sigma_{r=1}^{\infty} f(r\varepsilon(0)) w^r/m^r r!) \end{bmatrix} .$$

Proof: ρ clearly is one-to-one and functorial in R. Since $U'=\{\varepsilon(1)\}$, $\rho_R(\delta)$ is the identical transformation of $R[[w,z]]$. A straight-forward calculation shows that for $f,g \in \mathrm{Mun}(R)$, $\rho_R(f) \circ \rho_R(g) = \rho_R(fg)$. □

In particular,

$$\rho(\zeta) = \begin{bmatrix} w \to \exp(w)-1 \\ z \to z \exp(w/m) \end{bmatrix} \quad \text{and} \quad \rho(D^{\dim}) = \begin{bmatrix} w \to \frac{\exp(Dw)-1}{D} \\ z \to z \exp(Dw/m) \end{bmatrix}$$

because $\dim(r\varepsilon(0))=r$ and $\dim(\varepsilon(n))=n-1$ by (1.35). Thus the functions ζ^2, $\mu=\zeta^{-1}$, $\chi=\mu D^{\dim}$ and $D^{\dim}\zeta$ are mapped to the transformations

$$\rho(\zeta^2) = \begin{bmatrix} w \to \exp(\exp(w)-1)-1 \\ z \to z \exp(\frac{\exp(w)-1+w}{m}) \end{bmatrix} , \quad \rho(\mu) = \begin{bmatrix} w \to \log(1+w) \\ z \to z \ (1+w)^{-1/m} \end{bmatrix} ,$$

$$\rho(\chi) = \begin{bmatrix} w \to \frac{(1+w)^D-1}{D} \\ z \to z \ (1+w)^{(D-1)/m} \end{bmatrix} , \quad \rho(D^{\dim}\zeta) = \begin{bmatrix} w \to \exp(\frac{\exp(Dw)-1}{D})-1 \\ z \to z \exp(\frac{1}{m}(\frac{\exp(Dw)-1}{D} + Dw)) \end{bmatrix} .$$

We now apply Theorem (1.37) to the lattices $Q_r(G) = Q_{\{1,..,r\}}(G)$ studied by Dowling. For this, recall that $Q_r(G)$ has the least element $\hat{0}=(\varepsilon)$ and the greatest element $\hat{1}=(\emptyset)$, and observe that the type of the injection $(\{1,..,r\},\varepsilon) \to (\{1,..,r\},\emptyset)$ is $r\varepsilon(0)$.
Since $\#Q_r(G) = \zeta^2(r\varepsilon(0))$, $1+ \Sigma_{r=1}^{\infty} \#Q_r(G) w^r/m^r r! = \exp(\frac{\exp(w)-1+w}{m})$.
Replacing w by mw gives the exponential generating function for the cardinalities of the Dowling lattices:

$$(4.33) \qquad 1 + \Sigma_{r=1}^{\infty} \#Q_r(G) \ w^r/r! = \exp(\frac{\exp(mw)-1}{m} + w) \ .$$

Differentiating yields the recursion formula

$$(4.34) \qquad \#Q_{r+1}(G) = 2\#Q_r(G) + \Sigma_{j=0}^{r-1} \binom{r}{j} m^{r-j} \#Q_j(G)$$

starting with $\#Q_0(G) = 1$:

$$\#Q_1(G) = 2 \ , \ \#Q_2(G) = 4+m \ , \ \#Q_3(G) = m^2+6m+8 \ , \ \ldots \ .$$

The characteristic polynomial of $Q_r(G)$ is $\chi(r\varepsilon(0))$. To compute it, we develop $(1+w)^{(D-1)/m}$ into a binomial series and compare coefficients. It follows that

$$\chi(r\varepsilon(0)) = \Pi_{j=0}^{r-1} (D-1-jm) \quad ,$$

and, setting D=0, that

$$\mu_{Q_r(G)}(\hat{0},\hat{1}) = (-1)^r \, \Pi_{j=0}^{r-1} (1+jm) \quad .$$

In [Dow,73'] Dowling shows that the rank of $(\alpha) \in Q_r(G)$ is $r-\#\alpha$. Hence the Whitney numbers (of the second kind) of $Q_r(G)$ are given by the formula

$$W(i) = \Sigma_{j=0}^{i} \binom{r}{j} S(r-j,r-i) \, m^{i-j} \quad , \quad i=0,..,r \quad ,$$

which involves binomial coefficients and Stirling numbers of the second kind.

For finite geometric lattices it is conjectured that the Whitney numbers W(0),W(1),..,W(r) form a unimodal sequence ([We,76],p.288). We now verify this conjecture for the Dowling lattices by generalizing the proof for the partition lattices given in [Co,74],p.271. Instead of the polynomials $D^{\dim}\zeta(r\varepsilon(0)) = \Sigma_{i=0}^{r} W(i)D^i$ it is more convenient to use the reciprocal polynomials

$$P_r(D) = \Sigma_{i=0}^{r} W(i) D^{r-i}$$

because their generating function

$$\Phi(D,w) = \Sigma_{r=0}^{\infty} P_r(D) \, w^r/r! = \exp(D \, \frac{\exp(mw)-1}{m} + w)$$

is simpler than $\rho(D^{\dim}\zeta)(z)$. From $\partial\Phi/\partial w = (D+1)\Phi + mD\partial\Phi/\partial D$ we obtain the recursion formula

(4.35) $$P_{r+1} = (D+1)P_r + mDP_r' \quad , \quad P_0 = 1$$

where ' denotes differentiation. For example,

$P_1 = D+1$, $P_2 = D^2+(m+2)D+1$, $P_3 = D^3+(3m+3)D^2+(m^2+3m+3)D+1$.

Now consider the real functions $f_r(x) = (-x)^{1/m} e^x P_r(mx)/m^r$ defined for x<0. These functions approach 0 when x tends to $-\infty$ or to 0, and they satisfy the recursion $f_{r+1} = xf_r'$, $f_0(x) = (-x)^{1/m} e^x$. Applying the theorem of Rolle repeatedly shows that f_r has r distinct negative roots. From an inequality of Newton ([HLP,34],p.52) it follows that

$$W(i)^2 \geq \frac{(i+1)(r-i+1)}{i(r-i)} W(i-1)W(i+1) \qquad \text{for} \quad i=1,..,r-1 \ .$$

In particular, the sequence W(0),W(1),..,W(r) is logarithmically concave and thus unimodal.

APPENDIX

ALGEBRAIC REQUISITES

§1. Linear topological modules (compare [Kö,60],pp.85)

Let k be a commutative ring (associative with unit). We consider k as a topological ring with the discrete topology.
We call a Hausdorff topological k-module X linear topological if X has a basis of neighborhoods of zero (called a 0-basis) consisting of submodules. The standard example is k^S, S an arbitrary set, with the componentwise linear operations and the product topology.

Let X be a complete linear topological k-module. If $X = Y \oplus Z$ is a topological direct sum decomposition of X, we write $X = Y \hat{\oplus} Z$ and call Z a topological complement of Y in X. Given a family $(y_j; j\in J)$ of elements of X, the series $\Sigma_{j\in J} y_j$ is defined as the limit of the net $(\Sigma_{j\in F} y_j;$ F a finite subset of J) if the net converges. In particular, if (y_j) is a 0-family in X, i.e. if for every neighborhood N of 0 there are only finitely many $j\in J$ such that y_j lies outside N, then the series $\Sigma_{j\in J} y_j$ exists. For instance, if (y_n) is a sequence in X converging to 0, then also the series $\Sigma_{n=0}^{\infty} y_n$ converges.
A family $(x_i; i\in I)$ of elements of X is called a topological basis of X if the map $k^I \to X$, $(r_i; i\in I) \to \Sigma_{i\in I} r_i x_i$, is defined and a topological isomorphism. If Y_j, $j\in J$, are submodules of X such that the map $\Pi_{j\in J} Y_j \to X$, $(y_j; j\in J) \to \Sigma_{j\in J} y_j$, is defined and a topological isomorphism, we also write

$$X = \Pi_{j\in J} Y_j \quad .$$

Given two complete linear topological k-modules X and Y, the completed tensor product $X \hat{\otimes}_k Y = X \hat{\otimes} Y$ of X,Y is the Hausdorff completion of $X \otimes_k Y$ with respect to the topology given by the 0-basis $X_1 \otimes Y + X \otimes Y_1$ where X_1, Y_1 run over a 0-basis of X or Y respectively. For base ring extension, let Y=R be a commutative k-algebra endowed with the discrete topology. Then $X \hat{\otimes}_k R$ is a complete linear topological R-module. If in particular $X=k^S$, we have $k^S \hat{\otimes}_k R = R^S$ by an obvious identification.

Finally, we review the duality between the category $\underline{C}$ of topologically free topological k-modules and the category $\underline{D}$ of free discrete k-modules: If X is a topological k-module with the topological basis $(x_i; i \in I)$, then the k-module X' of all continuous linear functions from X to k is a discrete k-module with the dual basis $(x_i'; i \in I)$ where $x_i'(x_j) = \delta_{ij}$ for $i,j \in I$. Conversely, let U be a discrete k-module with basis $(u_j; j \in J)$. With the weak topology, the algebraic dual U* is a topological k-module with the topological basis $(u_j^*; j \in J)$ where $u_j^*(u_i) = \delta_{ji}$ for $j,i \in J$. This defines a duality, i.e. a contravariant category equivalence,

$$\text{(A.1)} \qquad \underline{C} \rightleftarrows \underline{D}\,, \qquad \begin{matrix} X & \to & X' \\ U^* & \leftarrow & U \end{matrix} \quad .$$

§2. Abstract incidence algebras

Let k be a commutative ring.

(A.2) DEFINITION: An abstract incidence algebra (AIA) over k is an associative topological k-algebra H with unit 1, together with a closed subalgebra $H^{(0)}$ and a filtration

$$H = H(0) \supset H(1) \supset H(2) \supset \ldots$$

of closed two-sided ideals of H, such that the following conditions are satisfied:

(a) H is Hausdorff, complete and has a 0-basis of two-sided ideals.

(b) As a topological k-algebra, $H^{(0)}$ is isomorphic to the product algebra k^S for some set S.

(c) (i) The filtration $H \supset H(1) \supset H(2) \supset ..$ converges to $\{0\}$, i.e. for any neighborhood N of 0 there is a number d such that $H(d) \subset N$.

(ii) For all $d_1, d_2 \in \mathbb{N}_o$, $H(d_1)H(d_2) \subset H(d_1+d_2)$.

(iii) For $d=1,2,..$, $H(d+1)$ is topologically complemented in $H(d)$ and $H(d)/H(d+1)$ is topologically free.

(d) H is the topological direct sum of $H^{(0)}$ and $H(1)$.

In the sequel, H is an AIA over k. The structure introduced above has many consequences.

The order of elements:

Since H is Hausdorff and $H \supset H(1) \supset H(2) \supset ..$ converges to $\{0\}$, we can define the order of an element $x \neq 0$ of H by

(A.3) $\quad \mathrm{ord}(x) = \max\{d \in \mathbb{N}_o;\ x \in H(d)\}$.

For convenience, we set $\mathrm{ord}(0) = \infty$. From condition (ii) we see that, for $x,y \in H$, $\mathrm{ord}(x+y) \geq \min(\mathrm{ord}(x),\mathrm{ord}(y))$ and $\mathrm{ord}(xy) \geq \mathrm{ord}(x)+\mathrm{ord}(y)$. Conversely, the order determines the filtration:

$$H(d) = \{x \in H;\ \mathrm{ord}(x) \geq d\} \quad , \quad d=0,1,2,.. \quad .$$

<u>Topological nilpotence and substitution into power series:</u>

The conditions (i),(ii) imply that, for any neighborhood N of 0, there is a d such that $H(1)^d \subset N$. Hence all elements of H(1) are <u>topologically nilpotent</u>, i.e. $\lim_d x^d = 0$ for all $x \in H(1)$. Thus, if $g = \Sigma_{d=0}^{\infty} g(d)X^d$ is a formal power series in the variable X,

(A.4) $$g(x) = \Sigma_{d=0}^{\infty} g(d)x^d$$

exists. In particular,

$$(1+x)^{-1} = \Sigma_{d=0}^{\infty} (-1)^d x^d \in 1+H(1) \quad .$$

This shows that H(1) is contained in the Jacobson radical of H. Therefore, an element y is invertible in H if and only if y+H(1) is invertible in H/H(1). But the decomposition $H = H^{(0)} \hat{\oplus} H(1)$, $y = p + x$, from condition (d) induces the isomorphism $H/H(1) \to H^{(0)}$, $y+H(1) \to p$. Thus we have proved

(A.5) PROPOSITION: For $y \in H$, let $y = p + x$ be the decomposition of y according to $H = H^{(0)} \hat{\oplus} H(1)$. Then y is invertible in H if and only if p is invertible in $H^{(0)}$. □

<u>Topological bases, multiplication constants and grading:</u>

For d=1,2,.. we have, by condition (iii), a decomposition

(A.6) $$H(d) = H^{(d)} \hat{\oplus} H(d+1)$$

where $H^{(d)}$ is a closed submodule of H. Note that in general this decomposition is not unique. Since H(d)/H(d+1) is topologically free and isomorphic to $H^{(d)}$, also $H^{(d)}$ is topologically free. Let $(e(t);t \in T^{(d)})$ be a topological basis of $H^{(d)}$. Obviously, $\mathrm{ord}(e(t))=d$ for all $t \in T^{(d)}$. As the filtration $H \supset H(1) \supset H(2) \supset ..$ converges to {0}, the map $\Pi_{d\geq 0} H^{(d)} \to H$, $(y^{(d)};d\geq 0) \to \Sigma_{d\geq 0} y^{(d)}$, is defined and a topological isomorphism, hence

$$H = \Pi_{d\geq 0} H^{(d)}$$

has the topological basis

(A.7) $$(e(t);t \in T) \quad \text{where} \quad T = \bigcup_{d\geq 0} T^{(d)} \quad .$$

Similarly, for l=1,2,.. , $H(l) = \Pi_{d\geq l} H^{(d)}$ has the topological basis

$(e(t);t\in T(d))$ where $T(d) = \dot{\cup}_{l\geq d} T^{(l)}$.

With respect to the fixed basis $(e(t);t\in T)$ of H, one obtains <u>multiplication constants</u> $G(t;t_1,..,t_l) \in k$ $(l\geq 1;\ t,t_1,..,t_l\in T)$ by the formula

(A.8) $$e(t_1)..e(t_l) = \Sigma_{t\in T} G(t;t_1,..,t_l)e(t) \quad .$$

In particular, the $G(t;t_1,t_2)$ are the <u>structure constants</u> of H. Finally, we call H <u>graded</u> by the sequence $(H^{(d)})_{d\geq 0}$ if

(A.9) $$H^{(d_1)}H^{(d_2)} \subset H^{(d_1+d_2)} \qquad \text{for all } d_1,d_2\in \mathbb{N}_o \quad .$$

(A.10) <u>The power series algebra:</u>
Let I be an arbitrary index set and let

$$P_k = k[[z_i;i\in I]]$$

be the k-algebra of formal power series in the indeterminates z_i. With the product topology, P_k is a topological algebra. We set $P_k^{(0)} = k1$ and define a filtration $P_k=P_k(0) \supset P_k(1) \supset P_k(2) \supset \ldots$ of P_k by the order of formal power series:

$$P_k(d) = \{a\in P_k;\ \text{order of } a \geq d\} \quad , \quad d=0,1,2,.. \quad .$$

With these data, P_k becomes an abstract incidence algebra over k. The order of an element $a\in P_k$ from (A.3) is just the order of the power series a. The standard grading of P_k is given by

$$P_k^{(d)} = \{a\in P_k;\ a \text{ homogeneous of degree } d\} \quad , \quad d=0,1,2,.. \quad .$$

The topological standard basis of $P_k^{(d)}$ are the monomials of degree d with coefficient 1. For power series algebras in finitely many indeterminates $z_1,..,z_r$ the change of the basis $z^n = z_1^{n_1}..z_r^{n_r}$ to another basis $y^n = y_1^{n_1}..y_r^{n_r}$ is interesting and treated by means of the Lagrange inversion formula, due to Abhyankar (compare [Joni,78] and [BCW,82]). Roman and Rota ([RR,78],p.132) call this the "transfer formula" and demonstrate its combinatorial usefulness (for the case of one variable). This is the reason why we require only the existence of a topological basis $(e(t);t\in T)$, but do not incorporate a distinguished one into the structure of an AIA.

(A.11) <u>The topological word algebra</u> ([Lo,82],pp.11; [Se,65],LA 4.13):
Let I be an arbitrary set, called alphabet, let Wo(I) be the free monoid of words over I and let

$$k\langle\langle I\rangle\rangle = \hat{Ass}_I$$

be the topological word algebra over I. In the usual way, we consider words as monomials in $k\langle\langle I\rangle\rangle$. With the subalgebra $k\langle\langle I\rangle\rangle^{(0)} = k1$ and the filtration induced by the standard grading

$$k\langle\langle I\rangle\rangle^{(d)} = \Pi\ \{kw;\ w \text{ a word of length } d\} \quad , \quad d=0,1,2,.. \quad ,$$

k<<I>> is an AIA over k. Given a finite set I, k<<I>> has the following universal property: If H is an AIA and if $y(i)$, $i \in I$, are arbitrary elements in $H(1)$, then there is a unique continuous k-algebra homomorphism φ: k<<I>> $\to$ H mapping each $i \in I$ to $y(i) \in H$, namely $\Sigma\, f(i_1,\ldots,i_n) i_1 \ldots i_n \to \Sigma\, f(i_1,\ldots,i_n) y(i_1) \ldots y(i_n)$.

Power series calculations:

We want to compute the expression $g(x)$ in formula (A.4) in a more explicit way. For this, some preparations are necessary.

Fix a topological basis $(e(t); t \in T)$ of H obtained in the non-unique way of (A.7). Let $G(t;t_1,\ldots,t_l) \in k$ be the multiplication constants of H with respect to this basis. For $(t_1,\ldots,t_l) \in T(1)^l$, we call

$$\gamma \in \mathbb{N}_o(T(1)) \quad , \quad \gamma(t) = \#\{i;\ t_i = t\} ,$$

the multiplicity of $(t_1,\ldots,t_l)$. Setting $|\gamma| = \Sigma_{t \in T(1)}\, \gamma(t)$, we have $|\gamma| = l$. Given $t \in T$ and $\gamma \in \mathbb{N}_o(T(1))$ with $|\gamma| = l$, we define

$$\text{(A.12)} \qquad G(t;\gamma) = \Sigma\, G(t;t_1,\ldots,t_l)$$

where the summation is taken over all $(t_1,\ldots,t_l) \in T(1)^l$ with multiplicity γ. Finally, for $t \in T$ and $l \geq 1$, we set

$$\text{(A.13)} \qquad G(t;l) = \Sigma_{t_1,\ldots,t_l \in T(1)}\, G(t;t_1,\ldots,t_l) = \Sigma_{|\gamma|=l}\, G(t;\gamma) \quad .$$

Obviously, $G(t;1) = 1$ if $t \in T(1)$ and $G(t;1) = 0$ otherwise .

Now we can describe the substitution into power series more explicitly.

(A.14) PROPOSITION: Let $x = \Sigma_{t \in T(1)}\, \lambda(t) e(t)$ be the representation of $x \in H(1)$ in the topological basis $(e(t); t \in T)$, and let $g = \Sigma_{d=0}^{\infty}\, g(d) X^d$ be a formal power series in the variable X. Then

$$g(x) = g(0)1 + \Sigma_{t \in T(1)}\, (\Sigma_{\gamma \neq 0}\, G(t;\gamma) g(|\gamma|) \lambda^{\gamma})\, e(t) \quad ,$$

where $\lambda^{\gamma} = \Pi_{v \in T}\, \lambda(v)^{\gamma(v)}$. For instance,

$$(1+x)^{-1} = 1 + \Sigma_{t \in T(1)}\, (\Sigma_{\gamma \neq 0}\, G(t;\gamma)(-1)^{|\gamma|} \lambda^{\gamma})\, e(t) \quad .$$

In particular,

$$(1 + \Sigma_{t \in T(1)}\, e(t))^{-1} = 1 + \Sigma_{t \in T(1)}\, (\Sigma_{l=1}^{\infty} (-1)^l G(t;l))\, e(t) \quad .$$

Proof: Since $e(t_1) \ldots e(t_d) = \Sigma_t\, G(t;t_1,\ldots,t_d) e(t)$,

$$\begin{aligned} x^d &= \Sigma_{t_1,\ldots,t_d \in T(1)}\, \lambda(t_1) \ldots \lambda(t_d) e(t_1) \ldots e(t_d) = \\ &= \Sigma_{t \in T(1)}\, (\Sigma_{t_1,\ldots,t_d \in T(1)}\, G(t;t_1,\ldots,t_d) \lambda(t_1) \ldots \lambda(t_d))\, e(t) = \\ &= \Sigma_{t \in T(1)}\, (\Sigma_{\gamma, |\gamma|=d}\, G(t;\gamma) \lambda^{\gamma})\, e(t) \qquad \text{and} \end{aligned}$$

$$g(x) = \Sigma_{d=0}^{\infty}\, g(d) x^d = g(0)1 + \Sigma_{t \in T(1)}\, (\Sigma_{\gamma \neq 0}\, G(t;\gamma) g(|\gamma|) \lambda^{\gamma})\, e(t) \quad . \ \square$$

<u>The group of invertible elements of H:</u>
For an arbitrary ring B, we denote the group of invertible elements of B by E(B). With the topology induced from H, E(H) is a Hausdorff topological group.

(A.15) PROPOSITION: E(H) is the semidirect product of the closed subgroup $E(H^{(0)})$ and the closed normal subgroup 1+H(1).

Proof: We construct the inverse of the map
$E(H^{(0)})\times(1+H(1)) \to E(H)$, $(b,c) \to bc$. Given $a\in E(H)$, write $a=p+y$ according to the decomposition $H=H^{(0)}\hat{\oplus}H(1)$. By Proposition (A.5) p is invertible, hence $a=p(1+p^{-1}y)$. Obviously, $(p,1+p^{-1}y) \leftarrow a$ is also continuous. □

Since by condition (b) $E(H^{(0)})$ is isomorphic to a product group $E(k)^S$, the algebraically interesting part of E(H) is the normal subgroup 1+H(1) which inherits the properties of H.

(A.16) COROLLARY: With the induced topology, the group 1+H(1) is a Hausdorff, complete topological group with a basis of neighborhoods consisting of normal subgroups. The filtration of 1+H(1)

$$1+H(1) \supset 1+H(2) \supset 1+H(3) \supset \dots$$

by closed normal subgroups converges to $\{1\}$, i.e. for any neighborhood N of 1 there is a number d such that $1+H(d) \subset N$. In particular, 1+H(1) is the inverse limit of the topological groups $1+H(1)/1+H(d)$, $d\geq 1$, with respect to the canonical projections
p_{nl}: $1+H(1)/1+H(l) \to 1+H(1)/1+H(n)$, $n\leq l$. □

(A.17) LEMMA: Let $(e(t);t\in T)$ be a topological basis of H as derived in (A.7). Then, for $1\leq n\leq l$, the maps

$$\alpha_{nl}:\ k^{T(n,l)} \to 1+H(n)/1+H(l) \ ,\ r \to \overline{1+\Sigma_{t\in T(n,l)} r(t)e(t)} \ ,$$

where $T(n,l) = T(n)-T(l) = \bigcup_{d=n}^{l-1} T^{(d)}$ and the bar indicates the coset, are homeomorphisms. If $l=n+1$, α_{nl} is even an isomorphism of topological groups.

Proof: The homeomorphism $H(n)\to 1+H(n)$, $x\mapsto 1+x$, induces the homeomorphism $\gamma_{nl}: H(n)/H(l) \to 1+H(n)/1+H(l)$, $\overline{x}\mapsto\overline{1+x}$. But $H(n)/H(l)$ is isomorphic to $\prod_{d=n}^{l-1} H^{(d)}$ which has the topological basis $(e(t);t\in T(n,l))$. Thus we get an isomorphism of topological k-modules

$k^{T(n,l)} \to H(n)/H(l)$, $r \to \overline{\Sigma_{t\in T(n,l)} r(t)e(t)}$.
If $l=n+1$, $H(n)H(n)\subset H(2n)\subset H(n+1)$ implies that $\gamma_{nl}(\bar{x})\gamma_{nl}(\bar{y})=(\overline{1+x})(\overline{1+y})=$ $=\overline{1+x+y}=\gamma_{nl}(\overline{x+y})$ for all $x,y\in H(n)$, thus γ_{nl} is even a group isomorphism.
□

(A.18) COROLLARY: For $d\geq 1$, the sequence of topological groups

$$0 \longrightarrow k^{T(d)} \xrightarrow{\beta_{d,d+1}} 1+H(1)/1+H(d+1) \xrightarrow{p_{d,d+1}} 1+H(1)/1+H(d) \longrightarrow 1$$

is exact. Here $\beta_{d,d+1}$ is the composition of $\alpha_{d,d+1}$ with the injection $1+H(d)/1+H(d+1) \to 1+H(1)/1+H(d+1)$. Therefore, the groups $1+H(1)/1+H(d)$ have normal series with factors of type k^I, I a set. □

The Lie algebra H(1):
Let L be a topological Lie algebra over the ring k and assume that L has a 0-basis of ideals. We call L topologically nilpotent if the descending central series

$$L=C^1(L) \supset C^2(L) \supset C^3(L) \supset \dots \qquad \text{(see [Se,65],LA 5.3)}$$

converges to $\{0\}$. The standard example for this notion is the topologically free Lie algebra on r generators

$$\hat{L}_I \subset \hat{Ass}_I \quad , \quad I=\{X_1,\dots,X_r\} \qquad \text{([Se,65],LA 4.13)}$$

which has the following universal property: If L is a topologically nilpotent Lie algebra and if $y_1,\dots,y_r$ are elements in L, then there is a unique Lie algebra homomorphism $\varphi\colon \hat{L}_I \to L$ mapping each X_i to y_i. If k contains the rationals and if $z(X,Y)\in \hat{L}_{\{X,Y\}}$ is the Lie power series from the Campbell-Hausdorff formula

$$\exp(X)\exp(Y) = \exp(z(X,Y)) \qquad \text{([Se,65],LA 4.15)} ,$$

then, for any $x,y\in L$, $z(x,y) = \varphi(z(X,Y))$ exists. Hence L is also a topological group with the multiplication $(x,y) \to z(x,y)$. We now apply this to the ideal H(1) of an AIA H.

(A.19) PROPOSITION: With the bracket $[x,y] = xy-yx$, H(1) is a topologically nilpotent Lie algebra. If k contains the rationals, the topological group 1+H(1) is isomorphic to H(1) with Campbell-Hausdorff multiplication by

$$H(1) \rightleftarrows 1+H(1) , \qquad \begin{matrix} x & \to & \exp(x) \\ \log(1+y) & \leftarrow & 1+y \end{matrix}$$

where $\exp(x)=\Sigma_{n=0}^{\infty} x^n/n!$ and $\log(1+y)=\Sigma_{n=1}^{\infty}(-1)^{n-1}/n\ y^n$.

Proof: By condition (i) it suffices to show that $C^n(H(1))\subset H(n)$ for all n. We proceed by induction on n: $C^1(H(1))=H(1)$ and, for $n\geq 2$, $C^n(H(1))=[H(1),C^{n-1}(H(1))]\subset[H(1),H(n-1)]\subset H(n)$. □

(A.20) Base ring extension:
If $H = \Pi\{ke(t);t\in T\}$ is an AIA over k and if R is a commutative k-algebra, then $H\hat{\otimes}_k R = \Pi\{Re(t);t\in T\}$ is an AIA over R. In particular, H defines a group functor

$$\Omega\colon \underline{Al}_k \to \underline{Gr}\ ,\ R \to E(H\hat{\otimes}_k R)\quad .$$

§3. Affine group schemes (see [DG,70],Chapter II; and [Sw,69])

Let k be a commutative ring. We denote the category of commutative k-algebras (associative with unit) by $\underline{Al}_k$, the category of monoids by $\underline{Mo}$ and the category of groups by $\underline{Gr}$.
An affine monoid (-scheme, -functor) over k is a representable functor

$$G\colon \underline{Al}_k \to \underline{Mo}\quad .$$

If $F_R\colon Al_k(A,R) \to G(R)$, $R\in Al_k$, is a functorial isomorphism, then A is the affine algebra of G and $x=F_A(id_A)$ is the corresponding universal element. F can be recovered from A and x by $F_R(f)=(Gf)(x)$, $f\colon A\to R$. With the comultiplication

$$\Delta\colon A \to A\otimes A\ ,\ a \to \Sigma\, a_{(1)}\otimes a_{(2)}\ ,$$

and the counit

$$\varepsilon\colon A \to k\quad ,$$

A is a commutative bialgebra (the contravariant bialgebra of G). We call G k-free if A is a free module over k. In the same fashion, one defines affine groups over k as representable group functors. In this case, the affine algebra A is a commutative Hopf algebra (the contravariant Hopf algebra of G) which in addition to Δ and ε has also a coinverse or antipode

$$S\colon A \to A\quad .$$

In the following let G be a k-free affine monoid. Without loss of generality we assume that $G = Al_k(A,-)$, i.e. $G(R) = Al_k(A,R)$ for $R\in Al_k$. Given an algebra $R\in Al_k$, the k-module $Hom_k(A,R)$ of k-linear maps from A to R becomes an associative R-algebra by the convolution multiplication

$$fg = \mu(f\otimes g)\Delta\quad ,\quad \text{i.e.}\quad (fg)(a) = \Sigma\, f(a_{(1)})g(a_{(2)})\quad .$$

Here μ is the multiplication in R. The unit of $Hom_k(A,R)$ is $\eta\varepsilon$ where $\eta\colon k\to R$ denotes the structure map of R. Then

$$G(R) \subset Hom_k(A,R)$$

is a multiplicative submonoid. Furthermore, $Hom_k(A,R)$ is a complete linear topological R-module with the weak topology induced from R^A. In particular, consider the dual module $A^* = Hom_k(A,k)$ of A. Using a basis $(u_j;j\in J)$ in A, we get an isomorphism of topological k-algebras

$$\mathrm{Hom}_k(A,R) \to A^*\hat{\otimes}_k R \ , \ f \to \Sigma_{j\in J}\ f(u_j)u_j^* \ .$$

Here $(u_j^*;j\in J)$ is the dual topological basis of A*. The unit η and the multiplication μ of A induce by duality maps

$$\varepsilon=\eta^*:\ A^* \to k \quad \text{and} \quad \Delta=\mu^*:\ A^* \to A^*\hat{\otimes}A^*$$

such that A* becomes a cocommutative topologically k-free topological bialgebra (the <u>covariant bialgebra</u> of G). If G is an affine group, then A* is a Hopf algebra (the <u>covariant Hopf algebra</u> of G) with antipode

$$S=S^*:\ A^* \to A^* \ .$$

Finally, we consider the Lie algebra of an affine group $G=\mathrm{Al}_k(A,-)$. The algebra A* is also a topological Lie algebra with the bracket $[x,y] = xy-yx$. Recall that $\varepsilon:A\to k$ is the counit of A. An ε-derivation of A is a k-linear map $d:A \to k$ such that

$$d(ab) = \varepsilon(a)d(b) + \varepsilon(b)d(a) \qquad \text{for all } a,b\in A \ .$$

The set Lie(G) of all ε-derivations of A is a closed sub Lie algebra of A*, and is called the <u>Lie algebra</u> of G. Let A^+ denote the kernel of $\varepsilon:A\to k$. As a topological k-module, Lie(G) is isomorphic to $\mathrm{Hom}_k(A^+/(A^+)^2,k)$ by $d \to l$, $l(a+(A^+)^2) = d(a)$. Another method to compute Lie(G) uses dual numbers: Let $k[\tau]$ be the k-algebra of dual numbers and let pr denote the projection $k[\tau] \to k1$ according to the decomposition $k[\tau] = k1\oplus k\tau$. Then $G(pr):G(k[\tau]) \to G(k)$ is a group homomorphism, and

$$\mathrm{Lie}(G) \to \text{kernel of } G(pr) \quad , \quad d \to [a\to\varepsilon(a)+d(a)\tau] \ ,$$

is a bijection. If $g_1=\varepsilon+\tau_1 d_1$, $g_2=\varepsilon+\tau_2 d_2 \in G(k[\tau_1,\tau_2])$ where τ_1, τ_2 have vanishing squares, then simple computation shows that

$$g_1 g_2 g_1^{-1} g_2^{-1} = \varepsilon + \tau_1\tau_2[d_1,d_2] \quad .$$

Now, let H be an abstract incidence algebra over the ground ring k, and let $(e(t);t\in T)$ be a topological basis of H as derived in (A.7). The following proposition is an immediate consequence of the results in the previous section.

(A.21) PROPOSITION:

(i) Ω: <u>Al</u>$_k$ $\to$ <u>Gr</u> , $R \to E(H\hat{\otimes}_k R)$, is an affine group which is represented by the algebra $A = k[X(v),X(v)^{-1},X(w);\ v\in T^{(0)},w\in T(1)]$ through $\mathrm{Al}_k(A,R) \to \Omega(R)$, $f \to \Sigma_{t\in T}\ f(X(t))e(t)$, and has the universal element $\Sigma_{t\in T}\ X(t)e(t) \in \Omega(A)$.

(ii) K: <u>Al</u>$_k$ $\to$ <u>Gr</u> , $R \to E(H^{(0)}\hat{\otimes}_k R)$, and Λ: <u>Al</u>$_k$ $\to$ <u>Gr</u> , $R \to 1+H(1)\hat{\otimes}_k R$, are closed affine k-free subgroups of Ω defined by the relations $X(w)=0$ for all $w\in T(1)$, or $X(v)=1$ for all $v\in T^{(0)}$ respectively.

(iii) Ω is the semidirect product of the closed subgroup K and the closed normal subgroup Λ. □

By condition (b) in (A.2), $H^{(0)}\hat{\otimes}_k R$ is isomorphic to the product algebra R^S for some set S. Therefore, the affine group K is isomorphic to the product $G_m{}^S$ of the multiplicative group

$$G_m: \underline{Al}_k \to \underline{Gr} \ , \ R \to E(R) \ ,$$

and so is <u>diagonalizable</u>. Next we turn to the affine group Λ which is represented by the algebra $A(1) = k[X(w);w\in T(1)]$ through $Al_k(A(1),R) \to \Lambda(R)$, $h \to 1+ \Sigma_{w\in T(1)} h(X(w))e(w)$, and consider the filtration of Λ

$$\Lambda=\Lambda(1) \supset \Lambda(2) \supset \Lambda(3) \supset \ldots$$

by the closed normal subgroups

$$\Lambda(d): \underline{Al}_k \to \underline{Gr} \ , \ R \to 1+H(d)\hat{\otimes}_k R \ ,$$

which are defined by the relations $X(w)=0$ for all $w\in T(1,d)$ in A(1).

(A.22) PROPOSITION: The affine group Λ is isomorphic to the inverse limit of the affine groups $\Lambda/\Lambda(d)$ with respect to the canonical projections

$$p_{nl}: \Lambda/\Lambda(l) \to \Lambda/\Lambda(n) \quad , \quad n\leq l \quad .$$

Here the quotient is taken in the category of all group functors, is affine and the canonical map $can:\Lambda \to \Lambda/\Lambda(d)$ is faithfully flat.

Proof: From Lemma (A.17) we see that $\Lambda/\Lambda(d)$ is represented by the polynomial algebra $k[X(w);w\in T(1,d)]$ through $Al_k(k[X(w);w\in T(1,d)],R) \to \Lambda(R)/\Lambda(d)(R)$, $h \to \overline{1+ \Sigma_{w\in T(1,d)} h(X(w))e(w)}$. Under these representations, the $p_{nl}:\Lambda/\Lambda(l)\to\Lambda/\Lambda(n)$ correspond to the injections $k[X(w);w\in T(1,n)]\to k[X(w);w\in T(1,l)]$ of the affine algebras. □

By Corollary (A.18) we have the following result which shows that the group $\Lambda/\Lambda(d+1)$ is an extension of the group $\Lambda/\Lambda(d)$ by a product of the <u>additive group</u>

$$G_a: \underline{Al}_k \to \underline{Gr} \ , \ R \to R \text{ with } + \ .$$

(A.23) COROLLARY: For all $d\geq 1$, the sequence of affine groups

$$0 \longrightarrow G_a{}^{T(d)} \xrightarrow{\beta_{d,d+1}} \Lambda/\Lambda(d+1) \xrightarrow{p_{d,d+1}} \Lambda/\Lambda(d) \longrightarrow 1$$

with faithfully flat $p_{d,d+1}$ is exact. □

Now assume that <u>k is a field</u>. An affine group G is called <u>unipotent</u> if every nonzero linear representation of finite dimension has a nonzero

fixed vector ([DG,70],p.487). The standard example of a unipotent group is the additive group G_a. Recall the following stability theorem from [DG,70],p.485: Closed subgroups, projective limits and extensions of unipotent groups are again unipotent.
Thus the affine group Λ is unipotent. In the following theorem we summarize our results.

(A.24) THEOREM: Let H be an abstract incidence algebra over a field k. Then the affine group

$$\Omega\colon \underline{Al}_k \to \underline{Gr}\ ,\ R \to E(H\hat{\otimes}_k R)\ ,$$

of invertible elements is the semidirect product of the closed diagonalizable subgroup

$$K\colon \underline{Al}_k \to \underline{Gr}\ ,\ R \to E(H^{(0)}\hat{\otimes}_k R)\ ,$$

and the closed unipotent normal subgroup

$$\Lambda\colon \underline{Al}_k \to \underline{Gr}\ ,\ R \to 1+H(1)\hat{\otimes}_k R\ ,$$

and hence is <u>triangulable</u> ([DG,70],pp.491). □

Finally, we compute the Lie algebra of Λ using dual numbers. Here k can be a commutative ring. The kernel of
$\Lambda(pr)\colon\Lambda(k[\tau])=1+H(1)\hat{\otimes}_k k[\tau]\to\Lambda(k)=1+H(1)$
is $1+\tau H(1)$. For $1+\tau_1 x_1, 1+\tau_2 x_2 \in\Lambda(k[\tau_1,\tau_2])$, the commutator is
$(1+\tau_1 x_1)(1+\tau_2 x_2)(1-\tau_1 x_1)(1-\tau_2 x_2) = 1+\tau_1\tau_2(x_1x_2-x_2x_1)$.
Therefore, we identify Lie(Λ) with the Lie algebra H(1) from (A.19).

(A.25) COROLLARY: The Lie algebra of Λ, Lie(Λ)=H(1), is a topologically nilpotent Lie algebra. If the ring k contains the rationals, the power series exp and log in $H\hat{\otimes}_k R$ define inverse group isomorphisms

$$\mathrm{Lie}(\Lambda)\hat{\otimes}_k R \underset{\log}{\overset{\exp}{\rightleftarrows}} \Lambda(R)$$

where the group structure on the left is the Campbell-Hausdorff composition. □

§4. Power series representations of affine monoids

Let k be a commutative ring and let I be an arbitrary index set. Given a commutative k-algebra R, we know from (A.10) that the power series algebra

$$P_R = R[[z_i;i\in I]] = P_k\hat{\otimes}_k R$$

is an AIA over R. In the category of abstract incidence algebras over R, the morphisms are the continuous R-algebra homomorphisms preserving the distinguished subalgebra and the filtration. More precisely, a morphism from the AIA H to the AIA E is a continuous R-algebra map $\Phi: H \to E$ such that $\Phi(H^{(0)}) \subset E^{(0)}$ and $\Phi(H(d)) \subset E(d)$ for all d . In particular, the <u>endomorphisms</u> of P_R are the continuous R-algebra maps $P_R \to P_R$ which preserve the ideal $P_R(1)$ of all formal power series with zero constant term. For this, observe that $P_R(d)$ is the closure of $P_R(1)^d$. Denoting the set of endomorphisms of P_R by $\mathrm{End}(P_R)$, we obtain the monoid functor

$$\mathrm{End}(P): \underline{\mathrm{Al}}_k \to \underline{\mathrm{Mo}} \;, \; R \to \mathrm{End}(P_R) \;,$$

which in general is not representable. Similarly, one defines the <u>automorphism</u> group scheme

$$\mathrm{Aut}(P): \underline{\mathrm{Al}}_k \to \underline{\mathrm{Gr}} \;, \; R \to \mathrm{Aut}(P_R) \;.$$

(A.26) DEFINITION: Let G be an affine monoid over k. A <u>power series representation</u> of G is a homomorphism $\rho: G \to \mathrm{End}(P)$. If G is an affine group, then $\rho: G \to \mathrm{Aut}(P)$. We call ρ <u>faithful</u> if all ρ_R, $R \in \mathrm{Al}_k$, are one-to-one. The power series representations $\rho, \sigma: G \to \mathrm{End}(P)$ are <u>equivalent</u> if there is an automorphism φ of P_k such that the diagram

$$\begin{array}{ccc} & \overset{\rho}{\nearrow} & \mathrm{End}(P) \\ G & & \big\downarrow c(\varphi) \\ & \underset{\sigma}{\searrow} & \mathrm{End}(P) \end{array}$$

commutes. Here $c(\varphi)_R: \mathrm{End}(P_R) \to \mathrm{End}(P_R)$, $R \in \mathrm{Al}_k$, is the functorial isomorphism defined by conjugation with $\varphi \hat{\otimes}_k R \in \mathrm{Aut}(P_R)$, $(\varphi \hat{\otimes}_k R)(a \hat{\otimes} r) = r\varphi(a)$. In other words, ρ is equivalent to σ if ρ can be transformed into σ by a change of coordinates in P_k.

From now on, we shall deal only with power series representations in finitely many indeterminates although in this book also examples for the infinite case occur (in chapter IV, §2,5,6). Thus, let

$$P_R = R[[z_1, \ldots, z_r]] \quad .$$

The endomorphisms of P_R are just the R-algebra maps $P_R \to P_R$ preserving $P_R(1)$. But $P_R(1)^r$ is a monoid under componentwise substitution with unit $(z_1, \ldots, z_r)$, and the monoid functor

$$\underline{\mathrm{Al}}_k \to \underline{\mathrm{Mo}} \;, \; R \to P_R(1)^r \;,$$

is representable. Since

$$\mathrm{End}(P_R) \to P_R(1)^r \;, \; \Phi \to (\Phi(z_i))_{i=1}^r \;,$$

is an anti-isomorphism of monoids and functorial in $R \in \mathrm{Al}_k$, the monoid scheme End(P) and, by the inverse function theorem, also the group scheme Aut(P) is affine.

Next we compute the Lie algebra of Aut(P). The kernel of

$Aut(P)(pr):Aut(P_{k[\tau]}) \to Aut(P_k)$ consists of all automorphisms Φ of the form $\Phi(z_i)=z_i+D(z_i)$, $i=1,..,r$, where D is a derivation of P_k such that $D(P_k(1))\subset P_k(1)$. Recall that a derivation of a k-algebra C is a k-linear map $D:C\to C$ obeying the product rule $D(bc)=D(b)c+bD(c)$ for all $b,c\in C$, and that the set Der(C) of derivations of C is a Lie algebra with the bracket $[D_1,D_2]=D_1D_2-D_2D_1$. Since the commutator of $\Phi_1=Id+D_1\tau_1$ and $\Phi_2=Id+D_2\tau_2$ is given by $\Phi_1\Phi_2\Phi_1^{-1}\Phi_2^{-1}(z_i)=z_i+\tau_1\tau_2(D_1D_2-D_2D_1)(z_i)$, $i=1,..,r$, we can identify

$$Lie(Aut(P)) = \{D\in Der(P_k);\ D(P_k(1))\subset P_k(1)\} .$$

But $Der(P_k)$ is a free P_k-left module with basis $\partial/\partial z_1,..,\partial/\partial z_r$ (the partial derivatives), a $D\in Der(P_k)$ having the basis representation $D=\Sigma_{i=1}^r D(z_i)\partial/\partial z_i$. Hence

$$Lie(Aut(P)) = \oplus_{i=1}^r P_k(1)\partial/\partial z_i .$$

If G is an affine group, every power series representation $\rho:G \to Aut(P)$ induces a representation $Lie(\rho):Lie(G) \to Lie(Aut(P))$ of the Lie algebra of G by derivations of P_k. Using dual numbers, $Lie(\rho)$ can be computed by restricting $\rho_{k[\tau]}:G(k[\tau]) \to Aut(P_{k[\tau]})$ to the kernel of G(pr).

Finally, we review some notions. An endomorphism Ψ of P_R is called <u>linear</u> if there is an r×r-matrix (λ_{ij}) with entries in R such that $\Psi(z_i) = \Sigma_{j=1}^r \lambda_{ij}z_j$ for $j=1,..,r$. Ψ is <u>diagonal</u> if (λ_{ij}) is a diagonal matrix. We call an endomorphism of P_R <u>diagonalizable</u> if it is conjugate in $Aut(P_R)$ to a diagonal endomorphism. A linear operator S on P_R is <u>topologically nilpotent</u> if $\lim_{n\to\infty} S^n(a) = 0$ for every $a\in P_R$. We say that an automorphism Φ is <u>topologically unipotent</u> if Φ-Id is topologically nilpotent.

(A.27) PROPOSITION: Let D be a diagonalizable affine monoid over a field k, and let $\rho:D \to End(P)$ be a power series representation of D in finitely many variables. Then D acts by endomorphisms which are simultaneously diagonalizable.

Proof: Consider the linear representations of D on $P_k(1)/P_k(d)$ induced by ρ. Since D is diagonalizable, there are power series $y_{1,2},..,y_{r,2}$ in $P_k(1)$ whose images in $P_k(1)/P_k(2)$ form a k-basis such that, for every $g\in D(R)$, $gy_{i,2} \equiv \chi_i(g)y_{i,2}$ modulo $P_k(2)$, $i=1,..,r$. Here $\chi_1,..,\chi_r$ are characters of D. By the following Lemma, we can construct for each i inductively elements $y_{i,3},y_{i,4},..$ in $P_k(1)$ such that $y_{i,d} \equiv y_{i,d+1}$ modulo $P_k(d)$ and $gy_{i,d} \equiv \chi_i(g)y_{i,d}$ modulo $P_k(d)$ for every $g\in D(R)$. Then the sequence $(y_{i,d})_d$ converges to a power series y_i in $P_k(1)$, and $gy_i = \chi_i(g)y_i$ for every $g\in D(R)$. Since $y_i \equiv y_{i,2}$ modulo $P_k(2)$, the $y_1,..,y_r$ are a coordinate system in P_k, and conjugation by the automorphism φ, $\varphi(y_i)=z_i$, yields a representation where D acts by diagonal endomorphisms.

For the construction of the $y_{i,d}$, we have used a result from linear algebra:

LEMMA: In a vector space V of finite dimension over k, let $(f_\gamma;\gamma\in\Gamma)$ be a family of commuting diagonalizable linear operators on V which leave a subspace U invariant. Suppose that $w\in V/U$ is a common eigenvector of the induced operators on V/U. Then the f_γ have a common eigenvector $v\in V$ whose image in V/U is w. □

(A.28) THEOREM: Let G be a triangulable affine group over a field k which is the semidirect product of the closed diagonalizable subgroup D and the closed unipotent normal subgroup U. Let $\rho: G \to \mathrm{Aut}(P)$ be a power series representation of G in finitely many indeterminates. Then ρ is equivalent to a power series representation $\sigma: G \to \mathrm{Aut}(P)$ of the following shape: For $i=1,\ldots,r$,

$$\sigma_R(g)(z_i) \in Rz_i \qquad \text{if } g\in D(R)$$

and

$$\sigma_R(g)(z_i)-z_i \in Rz_1+\ldots+Rz_{i-1}+P_R(2) \qquad \text{if } g\in U(R) .$$

Hence the induced Lie algebra representation $\mathrm{Lie}(\sigma):\mathrm{Lie}(G) \to \mathrm{Lie}(\mathrm{Aut}(P))$ looks like this: For $i=1,\ldots,r$,

$$\mathrm{Lie}(\sigma)(d)(z_i) \in kz_i \qquad \text{if } d\in \mathrm{Lie}(D)$$

and

$$\mathrm{Lie}(\sigma)(d)(z_i) \in kz_1+\ldots+kz_{i-1}+P_k(2) \qquad \text{if } d\in \mathrm{Lie}(U) .$$

Proof: Choose an equivalent representation π of G where D acts by diagonal automorphisms, and consider the induced linear representation of G on $P_k(1)/P_k(2)$. Since G is triangulable, there is a linear basis $y_1,\ldots,y_r$ of $P_k^{(1)}$ such that, for $i=1,\ldots,r$, $\pi_R(g)(y_i)\in Ry_i$ if $g\in D(R)$ and $\pi_R(g)(y_i)-y_i\in Ry_1+\ldots+Ry_{i-1}+P_R(2)$ if $g\in U(R)$. Changing to the new coordinate system $y_1,\ldots,y_r$, we have found a suitable representation σ. □

(A.29) COROLLARY: In every power series representation in finitely many indeterminates over a field, a unipotent affine group acts by topologically unipotent automorphisms whereas its Lie algebra operates by topologically nilpotent derivations. □

REFERENCES

Aigner M.
[Ai,79] Combinatorial Theory, Springer, Berlin, 1979.

Andrews G.E.
[An,76] The Theory of Partitions, Addison-Wesley, 1976, Reading, Mass.

Atiyah M.
[At,56] On the Krull-Schmidt Theorem with Applications to Sheaves, Bull. Soc. Math. France 84 (1956), 307-317.

Barnabei M. - A.Brini - G.C.Rota
[BBR,80] Sistemi di Coefficienti Sezionali I, Rendiconti del Circolo Mathematico di Palermo, Serie II, Tomo XXIX (1980), 457-484.

Bass H. - E.H.Connell - D.Wright
[BCW,82] The Jacobian conjecture: reduction of degree and formal expansion of the inverse, Bull. Amer. Math. Soc. 7 (1982), 287-330.

Bender E.A. - J.R.Goldman
[BG,71] Enumerative Uses of Generating Functions, Indiana Univ. Math. J. 20 (1971), 753-765.
[BG,75] On the Applications of Möbius Inversion in Combinatorial Analysis, Am. Mathem. Monthly 82 (1975), 789-803.

Berge C.
[Be,71] Principles of Combinatorics, Academic Press, New York, 1971.

Bucur I. - A.Deleanu
[BD,68] Introduction to the Theory of Categories and Functors, Wiley, London, 1968.

Burnside W.
[Bur,55] Theory of Groups of Finite Order, 2nd ed. 1911, Dover Publications, 1955.

Butcher J.C.
[But,72] An Algebraic Theory of Integration Methods, Math. Computation 26 (1972), 79-106.

Cartan H. - S.Eilenberg
[CE,56] Homological Algebra, Princeton Univ. Press, 1956.

Cartier P. - D.Foata
[CF,69] Problèmes combinatoires de commutation et réarrangements, LN in Math. 85, Springer, 1969.

Chen W.K.
[Ch,76] Applied Graph Theory, North Holland, Amsterdam, 1976.

Chowla S. - I.N.Herstein - W.R.Scott
[CHS,52] The solutions of $x^d=1$ in symmetric groups, Norske Vid. Selsk. 25 (1952), 29-31.

Cigler J.
[Ci,82] Elementare q-Identitäten, Publ. IRMA, Strasbourg (1982), 23-57.

Comtet L.
[Co,74] Advanced Combinatorics, Reidel, Dodrecht, 1974.

Content M. - F.Lemay - P.Leroux
[CLL,80] Catégories de Möbius et fonctorialités: un cadre général pour l'inversion de Möbius, J. Combin. Theory Ser.A 28 (1980), 169-190.

Crapo H.H.
[Cr,66] The Möbius Function of a Lattice, J. of Combin. Theory 1 (1966), 126-131.
[Cr,68] Möbius inversion in lattices, Arch. Math. (Basel) 19 (1968), 595-607.

Crapo H.H. - G.C.Rota
[CR,71] Combinatorial Geometries, M.I.T. Press, Cambridge, Mass., 1971.

Delsarte S.
[Del,48] Fonctions de Möbius sur les groupes abeliens finis, Annals of Math. 49 (1948), 600-609.

Demazure M. - P.Gabriel
[DG,70] Groupes Algébriques, North Holland, Amsterdam, 1970.

Désarménien J.
[Des,80] La fonction de Möbius du monoide plaxique, C.R.Acad.Sci.Paris Ser.A-B 290, no.19 (1980), A859-A861.

Doubilet P.
[Dou,72] Symmetric Functions through the Theory of Distribution and Occupancy, Studies in Appl. Math. 51 (1972), 377-396.
[Dou,74] Studies in partitions and permutations, Ph.D. thesis, M.I.T., 1974.

Doubilet P. - G.C.Rota - R.Stanley
[DRS,72] The idea of generating function, Proc. 6th Berkeley Symp. on Math.Stat. and Prob. vol.II: Probability Theory, 267-318, Univ.Calif. (1972).

Dowling T.A.
[Dow,73] A q-analog of the partition lattice, in: "A Survey of Combinatorial Theory", North Holland, Amsterdam, 1973.
[Dow,73'] A class of geometric lattices based on finite groups, J. Combin. Theory Ser.B 14 (1973), 61-86.

Dür A.
[Dür,81] Inzidenzalgebren zu Cozyklen mit nichtleerer Nullstellenmenge, Diplomarbeit, Innsbruck, 1981.
[Dür,83] Unipotente Gruppen in der Kombinatorik, Dissertation, Innsbruck, 1983.

Dür A. - U.Oberst
[DO,82] Incidence Algebras, Exponential Formulas and Unipotent Groups, Proc., LN in Math. 969, 133-166, Springer, 1982.

Erdélyi A. - W.Magnus - F.Oberhettinger - F.G.Tricomi
[EMOT,55] Higher Transcendental Functions, vol.III, McGraw-Hill, New York, 1955.

Foata D.
[F,74] La série génératrice exponentielle dans les problèmes d'énumérations, Les Presses de l'Université de Montréal, 1974.

Foata D. - M.P.Schützenberger
[FS,70] Théorie Géométrique des Polynomes Eulériens, LN in Math. 138, Springer, 1970.

Garsia A.M. - J.Remmel
[GaRe,80] A Combinatorial Interpretation of q-Derangement and q-Laguerre Numbers, European J. Combinatorics 1 (1980), 47-59.

Goldman J. - G.C.Rota
[GR,70] Finite Vector Spaces and Eulerian Generating Functions, Studies in Appl. Math. 49 (1970), 239-258.

Goldschmidt H.
[Go,72] Sur la Structure des Equations de Lie: I. Le Troisième Théorème Fondamental, J. Differential Geometry 6 (1972), 357-373.

Graver J.E. - M.E.Watkins
[GW,77] Combinatorics with Emphasis on the Theory of Graphs, Springer, New York, 1977.

Greene C.
[Gr,82] The Möbius function of a partially ordered set, in: Ordered sets (Banff, Alta., 1981), Proc. of the NATO Advanced Study Institute, Reidel, Dordrecht, 1982, 555-581.

Greene C. - D.J.Kleitman
[GK,78] Proof techniques in the theory of finite sets, in: Studies in Combinatorics, MAA Studies in Math. 17 (1978), 22-79.

Grothendieck A. - J.V.Verdier
[GV,72] Theorie des Topos (SGA 4, exposés I-VI). Second edition, Springer LN in Math., 269-270 (1972).

Hairer E. - G.Wanner
[HW,74] On the Butcher group and general multivalue methods, Computing 13 (1974), 1-15.

Hall P.
[Ha,59] The algebra of partitions, Proc. of the Fourth Canadian Math. Congress (1957), Banff (1959), 147-159.

Hanlon P.
[Han,81] The Fixed-Point Partition Lattices, Pacific J. of Math. 96 (1981), 319-341.
[Han,84] The Characters of the Wreath Product Group Acting on the Homology Groups of the Dowling Lattices, J. of Algebra 91 (1984), 430-463.

Hardy G.H. - J.E.Littlewood - G.Pôlya
[HLP,34] Inequalities, Cambridge University Press, 1934.

Henle M.
[He,72] Dissection of Generating Functions, Studies in Appl. Math. 51 (1972), 397-410.
[He,75] Binomial enumeration on dissects, Trans. Amer. Math. Soc. 202 (1975), 1-38.

Hilton P.J. - U.Stammbach
[HS,71] A Course in Homological Algebra, Springer, New York, 1971.

James G. - A.Kerber
[JK,81] The Representation Theory of the Symmetric Group, Addison-Wesley, 1981, Reading, Mass.

Johnstone P.T.
[Joh,77] Topos Theory, Academic Press, London, 1977.

Joni S.A.
[Joni,78] Lagrange inversion in higher dimensions and umbral operators, Lin. Multil. Alg. 6 (1978), 111-122.

Joni S.A. - G.C.Rota
[JR,79] Coalgebras and Bialgebras in Combinatorics, Studies in Appl. Math. 61 (1979), 93-139.

Joyal A.
[Joy,81] Une théorie combinatoire des séries formelles, Advances in Math. 42 (1981), 1-82.

Kerber A. - K.J.Thürlings
[KT,82] Counting symmetry classes of functions by weight and automorphism group, Proc., LN in Math. 969, 191-211, Springer, 1982.
[KT,83] Symmetrieklassen von Funktionen und ihre Abzählungstheorie I und II, Bayreuther Mathematische Schriften 12 und 15 (1983).

Köthe G.
[Kö,60] Topologische Lineare Räume I, Springer, Berlin, 1960.

Kung J.P.S.
[Ku,81] The Cycle Structure of a Linear Transformation over a Finite Field, Linear Algebra and Its Appl. 36 (1981), 141-155.

Lascoux A. - M.P.Schützenberger
[LS,78] Le monoide plaxique, Quaderni della Ricerca Scientifica del C.N.R. (â paraitre), 1978.

Lindstrom B.
[Li,69] Determinants on semilattices, Proc. AMS 20 (1969), 207-208.

Lothaire M.
[Lo,82] Combinatorics on Words, Addison-Wesley, 1982, Reading, Mass.

Macdonald I.G.
[Ma,79] Symmetric Functions and Hall Polynomials, Clarendon Press, Oxford, 1979.

MacLane S.
[ML,71] Categories for the Working Mathematician, Springer, New York, 1971.

Nichols W. - M.Sweedler
[NS,80] Hopf algebras and combinatorics, in: Umbral Calculus and Hopf Algebras, Contemporary Mathematics vol.6, AMS, 1980.

Pavlov A.I.
[Pa,81] On the number of solutions of the equation $x^k=a$ in the symmetric group S_n, Math. USSR Sbornik 40 (1981), 349-362.

Riordan J.
[Ri,58] An Introduction to Combinatorial Analysis, Wiley, New York, 1958.
[Ri,68] Combinatorial Identities, Wiley, New York, 1968.

Roman S.M. - G.C.Rota
[RR,78] The Umbral Calculus, Advances in Math. 27 (1978), 95-188.

Roos J.E.
[Roos,61] Sur les foncteurs dêrivés de lim. Applications. C.R.Acad.Sci.Paris 252 (1961), 3702-3704.

Rosen J.
[Ros,76] The Number of Product-Weighted Lead Codes for Ballots and Its Relation to the Ursell Functions of the Linear Ising Model, J. of Combin. Theory Ser.A 20 (1976), 377-384.

Rota G.C.
[Ro,64] Theory of Möbius Functions, Z. Wahrscheinlichkeitstheorie 2 (1964), 340-368.

Schützenberger M.P.
[Sch,54] Contribution aux applications statistiques de la thêorie de l'information, Publ. Inst. Stat. Univ. Paris 3, 5-117 (1954).

Serre J.P.
[Se,65] Lie Algebras and Lie Groups, Benjamin, New York, 1965.

Stanley R.P.
[St,70] Structure of incidence algebras and their automorphism groups, AMS Bull. 76 (1970), 1236-1239.
[St,72] Supersolvable lattices, Algebra Universalis 2 (1972), 197-217.
[St,74] Combinatorial reciprocity theorems, Advances in Math. 14 (1974), 194-253.
[St,76] Binomial Posets, Möbius Inversion and Permutation Enumeration, J. Combin. Theory Ser.A 20 (1976), 336-356.
[St,78] Generating Functions, in: Studies in Combinatorics, MAA Studies in Math.17 (1978), 100-141.
[St,78'] Exponential Structures, Studies in Appl. Math. 59 (1978), 73-82.
[St,82] Some Aspects of Groups Acting on Finite Posets, J. Combin. Theory Ser.A 32 (1982), 132-161.

Sweedler M.E.
[Sw,69] Hopf Algebras, Benjamin, New York, 1969.

Sylvester G.S.
[Sy,76] Continuous-Spin Ising Ferromagnets, thesis, M.I.T., 1976.

Tutte W.T.
[Tu,70] Introduction to the Theory of Matroids, American Elsevier, New York, 1970.

Welsh D.J.A.
[We,76] Matroid Theory, Academic Press, London, 1976.

A LIST OF RECENT ARTICLES CONNECTED WITH COMBINATORIAL MÖBIUS INVERSION

Athreya K.B. - C.R.Pranesachar - N.M.Singhi
On the number of Latin rectangles and chromatic polynomials of $L(K_{r,s})$,
European J. Combinatorics 1, no.1 (1980), 9-17.

Baclawski K.
Cohen-Macaulay connectivity and geometric lattices,
European J. Combinatorics 3, no.4 (1982), 293-305.

Baclawski K. - A.M.Garsia
Combinatorial decomposition of a class of rings,
Advances in Math. 39 (1981), 155-184.

Barnabei M. - A.Brini
Some properties of characteristic polynomials and applications to T-lattices,
Discrete Math. 31, no.3 (1980), 261-270.

Björner A.
1. Homotopy type of posets and lattice complementation,
   J. Combin. Theory Ser.A 30, no.1 (1981), 90-100.
2. On the homology of geometric lattices,
   Algebra Universalis 14, no.1 (1982), 107-128.
3. Some combinatorial and algebraic properties of Coxeter complexes and Tits
   buildings, Advances in Math. 52 (1984), 173-212.

Björner A. - A.M.Garsia - R.P.Stanley
An introduction to Cohen-Macaulay partially ordered sets, in: Ordered sets (Banff, Alta., 1981), Proc. of the NATO Advanced Study Institute, Reidel, Dordrecht, 1982, 583-615.

Björner A. - M.Wachs
Bruhat order of Coxeter groups and shellability,
Advances in Math. 43, no.1 (1982), 87-100.

Björner A. - J.W.Walker
A homotopy complementation formula for partially ordered sets,
European J. Combinatorics 4, no.1 (1983), 11-19.

Brini A.
1. A class of rank-invariants for perfect matroid designs,
   European J. Combinatorics 1, no.1 (1980), 33-38.
2. Some homological properties of partially ordered sets,
   Advances in Math. 43 (1982), 197-201.

Bulgak A.S.
The Möbius function of the lattice of subgroups of a finite group,
Moskov.Inst.Inzh.Zheleznodorozh.Transporta Trudy No. 653 (1982), 103-109.

Buttsworth R.N.
An inclusion-exclusion transform, Ars Combin. 15 (1983), 279-300.

Cartier P.
Les arrangements d'hyperplans: un chapitre de géométrie combinatoire,
Bourbaki Seminar, vol. 1980/1981, LN in Math. 901, 1-22, Springer, 1981.

Cerasoli M.
1. The Möbius-Rota function as an enumeration method,
   Mathematics review (Italian), Tilgher, Genoa, 1980, 73-86.
2. On the lattice of the stones of a graph,
   J. Combin. Inform. System Sci. 5, no.2 (1980), 134-140.

Cohen D.I.A.
PIE-sums: a combinatorial tool for partition theory,
J. Combin. Theory Ser.A 31, no.3 (1981), 223-236.

Content M. - F.Lemay - P.Leroux
Catégories de Möbius et fonctorialités; un cadre général pour l'inversion de Möbius, J. Combin. Theory Ser.A 28 (1980), 169-190.

Cordovil R. - M. Las Vergnas - A.Mandel
Euler's relation, Möbius functions and matroid identities,
Geom. Dedicata 12, no.2 (1982), 147-162.

Désarménien J.
La fonction de Möbius du monoïde plaxique,
C.R.Acad.Sci.Paris Ser.A-B 290, no.19 (1980), A859-A861.

Dür A. - U.Oberst
Incidence algebras, exponential formulas and unipotent groups,
Proc., LN in Math. 969, 133-166, Springer, 1982.

Edelman P.H.
Zeta polynomials and the Möbius function,
European J. Combinatorics 1, no.4 (1980), 335-340.

Gluck D.
Idempotent formula for the Burnside algebra with applications to the p-subgroup simplicial complex, Illinois J. Math. 25 (1981), 63-67.

Greene C.
The Möbius function of a partially ordered set, in: Ordered sets (Banff, Alta., 1981), Proc. of the NATO Advanced Study Institute, Reidel, Dordrecht, 1982, 555-581.

Haigh J.
On the Möbius algebra and the Grothendieck ring of a finite category,
J. London Math. Soc. (2) 21, no.1 (1980), 81-92.

Hanlon P.
1. The incidence algebra of a group reduced partially ordered set, Combinatorial mathematics, VII (Proc. Seventh Australian Conf., Univ. Newcastle, Newcastle,1979), LN in Math. 829, 148-156, Springer, 1980.
2. The fixed-point partition lattices, Pacific J. of Math. 96 (1981), 319-341.
3. The characters of the wreath product group acting on the homology groups of the Dowling lattices, J. of Algebra 91 (1984), 430-463.

Harrison K.J. - W.E.Longstaff
Subalgebras of incidence algebras determined by equivalence relations,
J. Combin. Theory Ser.A 31 (1981), 94-97.

Hu Y.
Generalized inversion on posets and its applications,
J. Huazhong Univ. Sci. Tech. 11 (1983), no.5, 57-62.

Jambu M. - H.Terao
Free arrangements of hyperplanes and supersolvable lattices,
Advances in Math. 52 (1984), 248-258.

Joni S.A. - G.C.Rota
1. On restricted bases for finite fields, Annals Discrete Math. 6 (1980), 215-217.
2. A vector space analog of permutations with restricted position, J. Combin. Theory Ser.A 29 (1980), 59-73.

Joyal A.
Une théorie combinatoire des séries formelles, Advances in Math. 42 (1981), 1-82.

Kerber A. - K.J.Thürlings
1. Counting symmetry classes of functions by weight and automorphism group, Proc., LN in Math. 969, 191-211, Springer, 1982.
2. Symmetrieklassen von Funktionen und ihre Abzählungstheorie I,II und III, Bayreuther Mathematische Schriften 12(1983), 15(1983), und 21(1986),156-278.

Kratzer C. - J.Thévenaz
Fonction de Möbius d'un groupe fini et anneau de Burnside,
Comment. Math. Helvetici 59 (1984), 425-438.

Kreid T.
1. The theorem about inversion in generalized incidence algebras,
Demonstratio Math. 15 (1982), no.1, 181-187.
2. Application of polynomials to investigations of Möbius algebras,
Demonstratio Math. 15 (1982), no.3, 763-766.

Kung J.P.S.
The Redei function of a relation,
J. Combin. Theory Ser.A 29, no.3 (1980), 287-296.

Kung J.P.S. - M.R.Murty - G.C.Rota
On the Redei zeta function, J. Number Theory 12, no.3 (1980), 421-436.

Morris I. - C.B.Wensley
Adams operations and λ-operations in β-rings,
Discrete Math. 50 (1984), 253-270.

Mozzocca F.
Möbius functions and some applications of the inversion formula, Quaderni Serie III,
128, Istituto per le Appl. del Calcolo "Mauro Picone" (IAC), Rome, 1981, 23pp.

Narushima H.
Principle of inclusion-exclusion on partially ordered sets,
Discrete Math. 42 (1982), 243-250.

Nechvatal J.R.
Asymptotic enumeration of generalized Latin rectangles,
Utilitas Math. 20 (1981), 273-292.

Orlik P. - L.Solomon
1. Combinatorics and topology of complements of hyperplanes,
Invent. Math. 56, no.2 (1980), 167-189.
2. Complexes for reflection groups, in: Algebraic geometry (Chicago, 111., 1980),
LN in Math. 862 (1981), 193-207.

Palfy P.P.
On faithful irreducible representations of finite groups,
Studia Sci. Math. Hungar., 1979, 14, no.1-3, 95-98 (1982).

Plesken W.
Counting with groups and rings,
J. für reine und angewandte Mathematik 334 (1982), 40-68.

Rota G.C. - B.Sagan
Congruences derived from group action,
European J. Combinatorics 1, no.1 (1980), 67-76.

Sivaramakrishnan R.
On abstract Möbius inversion, Proc. Sec. Conf. Number Theory (Ootacamund,1980),
Matscience Rep. 104, 85-100, Inst. Math. Sci., Madras, 1981.

Speed T.B.
On the Möbius function of Hom(P,Q),
Bull. Austral. Math. Soc. 29 (1984), no.1, 39-46.

Speed T.P. - R.A.Bailey
On a class of association schemes derived from lattices of equivalence relations,
in: Algebraic structures and applications (Nedlands, 1980), LN in Pure and Appl.
Math. 74, 55-74, Dekker, New York, 1982.

Stanley R.P.
Some aspects of groups acting on finite posets,
J. Combin. Theory Ser.A 32 (1982), 132-161.

Stechkin B.S.
Imbedding theorems for Möbius functions,
Dokl.Akad.Nauk SSSR 260, no.1 (1981), 40-43.

Terao H.
On Betti numbers of complement of hyperplanes,
Publ. Res. Inst. Math. Sci. 17, no.2 (1981), 657-663.

Walker J.W.
Homotopy type and Euler characteristic of partially ordered sets,
European J. Combinatorics 2, no.4 (1981), 373-384.

Xu L.Z.
1. An extension of the Möbius-Rota inversion theory with applications,
J. Math. Res. Exposition, 1981, no.1, First Issue, 101-112.
2. Generalized Möbius-Rota inversion theory associated with nonstandard analysis,
Sci. Exploration 3 (1983), no.1, 1-8.

Xu L.Z. - J.X.Yang
On a method of constructing interpolation formulas via inverse series relations,
J. Math. Res. Exposition, 1982, 2, no.2, 113-126.

Yuzvinsky S.
Linear representations of posets, their cohomology and a bilinear form,
European J. Combinatorics 2, no.4 (1981), 385-397.

Zaslavsky T.
1. Signed graph coloring, Discrete Math. 39 (1982), 215-228.
2. Chromatic invariants of signed graphs, Discrete Math. 42 (1982), 287-312.

INDEX